Fernanda Krumreich
Ana Paula A. Correa
Jair Costa Nachtigal

Hydrocolloids and pectinase for stabilizing guava nectar

Fernanda Krumreich
Ana Paula A. Correa
Jair Costa Nachtigal

Hydrocolloids and pectinase for stabilizing guava nectar

Use of hydrocolloids and pectinase to stabilize guava nectar Psidium guajava L. cv. Paluma

ScienciaScripts

Imprint
Any brand names and product names mentioned in this book are subject to trademark, brand or patent protection and are trademarks or registered trademarks of their respective holders. The use of brand names, product names, common names, trade names, product descriptions etc. even without a particular marking in this work is in no way to be construed to mean that such names may be regarded as unrestricted in respect of trademark and brand protection legislation and could thus be used by anyone.

Cover image: www.ingimage.com

This book is a translation from the original published under ISBN 978-613-9-61346-5.

Publisher:
Sciencia Scripts
is a trademark of
Dodo Books Indian Ocean Ltd. and OmniScriptum S.R.L publishing group

120 High Road, East Finchley, London, N2 9ED, United Kingdom
Str. Armeneasca 28/1, office 1, Chisinau MD-2012, Republic of Moldova, Europe
Printed at: see last page
ISBN: 978-620-7-60893-5

Table of contents:

I dedicate this work to the most
important people in my life, my family.

Summary

KRUMREICH, Fernanda Döring. **Use of hydrocolloids and pectinase to stabilize guava nectar (*Psidium guajava* L. cv. Paluma)** 2015. 115f. Dissertation (Master's Degree) - Graduate Program in Food Science and Technology. Federal University of Pelotas, Pelotas.

The market for fruit juices and nectars has been growing in recent years, which is attributed to the concern about acquiring healthier and more beneficial foods. *Guava* (*Psidium guajava* L.), which belongs to the Myrtaceae family, is considered to be rich in bioactive compounds such as carotenoids, phenolic compounds and L-ascorbic acid. Its fruits are mainly consumed fresh or in the form of juices, nectars, jams and jellies. Nectar is one of the forms of consumption that has seen a significant increase, but it shows phase separation during storage. Based on this, the bioactive compounds of guava cv. Paluma pulp were characterized and nectars were formulated with the addition of hydrocolloids and/or enzymes in order to stabilize the nectar and evaluate its stability and that of its main bioactive compounds during storage for 180 days. The fruit was processed into pulp and then into nectars, totaling 11 treatments, including a control (pulp, water and sugar) and with the addition of xanthan, guar, pregelatinized rice flour and the enzyme pectinase, used alone and also in combination. The pulp was analyzed for bioactive compounds, physical and chemical parameters, color and mineral salts. For the nectar, in addition to the same assessments as for the pulp, except for mineral salts, sedimentation and apparent viscosity were assessed every 45 days of storage. The pulp proved to be a source of minerals, especially calcium, phosphorus, magnesium and potassium, with a high content of L-ascorbic acid and carotenoids. In general, the nectars showed a decrease in pH and an increase in total titratable acidity and soluble solids compared to the control. Throughout storage it was observed that the treatments with the enzyme pectinase showed a decrease in pH, an increase in the total titratable acidity content and an increase in the extraction of soluble solids and carotenoids, as well as less phase separation and less viscous samples, with the enzyme and guar treatment standing out among them. With regard to the highest antioxidant capacity, the treatment with xanthan stands out, as it provided an encapsulating solution for phenolic compounds and L-ascorbic acid (vitamin C), both of which are responsible for this capacity. It can thus be seen that both the pulp and the nectars had appreciable amounts of carotenoids and vitamin C, but there was a small decrease in carotenoids and large losses of vitamin C during the storage period of these nectars.

Keywords: bioactive compounds, nectar, storage, hydrocolloids, enzymes

Chapter 1

1. General introduction

The market for fruit juices and nectars has been growing significantly due to various reasons, such as the improvement of the products on offer, practicality, increased consumer income, high nutritional value, as well as concerns about acquiring healthier foods. As a result, the food industry is increasingly expanding its offer of quality drinks with a variety of flavors According to the Brazilian Association of Soft Drinks and Non-Alcoholic Beverages Industries (ABIR, 2011), in 2004, the volume of fruit juices and nectars sold in the country was 253 million liters, rising to 426 million liters in 2008, with guava juices and nectars accounting for 6.2% of this total. In its latest survey, in 2010, the sector sold 533 million liters, equivalent to 20% more than in 2008.

Fruits are foods that offer a wide variety of flavors and aromas, and are composed of high levels of water (around 80%) and carbohydrates (5 to 25%). They are an indispensable part of the human diet, being the main source of ascorbic acid, which is essential for the intake of carotene, riboflavin, thiamine, niacin, folic acid, as well as other B vitamins (MACHADO & LINHARES, 2002). They also contain fiber and many mineral salts, especially calcium, iron and phosphorus (ARTHEY & ASHURST, 1998).

Guava is a fruit rich in vitamin A and contains high levels of B vitamins, mainly thiamine (B1), riboflavin (B2) and niacin. Its levels of fiber, protein, total sugars and mineral elements such as calcium, phosphorus and potassium, as well as significant amounts of vitamin C, carotenoids and phenolic compounds, which act as antioxidants, capturing free radicals in the body, have anti-inflammatory, anti-platelet, anti-cancer and hypoglycemic effects, make this fruit one of the most complete and balanced in terms of nutritional value. In general, fruits are quickly digested and easily assimilated by the body. They can be consumed in their natural state or after being processed into juices, nectars, sweets, preserves or dried fruit (MANICA et al., 1998).

The acceptance of processed foods depends mainly on the food's ability to meet consumer expectations. Some sensory quality attributes such as taste, aroma and appearance determine this, along with other characteristics, the acceptance or rejection of these foods. Ready-to-drink juices and nectars are currently highly sought-after products, due to the growing consumer demand for more convenient, healthier and lower-calorie products (SUPERHIPER, 2000). In Brazil, there is still a need for more detailed studies on processing parameters and storage time on the quality of nectars made from some of the fruits produced in the country, which have the agronomic and varietal characteristics of tropical countries.

In the processing of guava nectars, the separation of phases that occurs shortly after production or during storage degrades the visual appearance of the product, thus compromising its competitiveness. One alternative to overcome this problem would be the clarification process; however, this process implies the partial removal of desirable components such as aromatic substances and natural antioxidants such as carotenoids. Alternatives are the use of hydrocolloids or gums in small quantities, which can keep the pulp suspended without appreciably altering its quality. By thickening the dispersant phase, a three-dimensional network is created which encompasses the particles, keeping them in suspension (GLICKSMAN, 1982).

With a view to stabilizing guava nectar, the aim of this study was to evaluate the efficiency of using hydrocolloids alone and in combination with the enzyme pectinase, as well as to assess physicochemical parameters, the content of bioactive compounds and antioxidant capacity during storage for 180 days at room temperature.

1.1 Hypotheses

The application of hydrocolloids in combined form is more effective in stabilizing guava nectar during storage than in isolated form.

The addition of the enzyme pectinase, in addition to its ability to stabilize guava nectar, also increases the retention of bioactive compounds.

1.2 Objectives

Characterize a *Psidium guajava* L. *guava* pulp in terms of physicochemical parameters, content of bioactive compounds and mineral salts.

To make guava nectar with the pulp of the Paluma cultivar, with the addition of the hydrocolloids xanthan, guar and pregelatinized rice flour and the enzyme pectinase, used alone and in combination.

To evaluate the stability of guava nectar processed and stored for 180 days, in terms of physicochemical parameters, content of bioactive compounds and antioxidant capacity.

Chapter 2

2. Literature review

2.1 The guava tree (*Psidium guajava* L.)

The guava tree belongs to the Myrtaceae family, genus Psidium, which currently comprises between 110 and 130 species, and of these, only the *Psidium guajava* L. *guava* tree has economic importance in different forms of exploitation (GONZAGA NETO; SOARES, 1995. JOSEPH & PRIYA, 2011). According to the Brazilian Institute of Geography and Statistics (IBGE), the amount of guava produced in the country in 2010 was 316,000 tons, with the Southeast and Northeast being the main producing regions, each accounting for almost 41% of national production. The state of São Paulo, the largest producer, accounted for 31% of this total, followed by Pernambuco with 29% (KIST et al., 2012). Countries such as France, the United Kingdom and the Netherlands are major consumers of Brazilian guava, together consuming almost 166,000 tons of the fruit, generating an income of approximately 310,000 dollars/year (FREITAS, 2010).

The guava tree is an evergreen shrub or semi-shrub, reaching a height of 3 to 7 meters. Its fruit varies in size, shape, taste and weight depending on the cultivar. The flesh can be white, cream, yellow, pink or red (TODA FRUTA, 2009).

Guava is one of the most important fruits in tropical and subtropical regions, not only because of its high nutritional value, but also because of its excellent acceptance for fresh consumption, its ability to thrive in adverse conditions and its wide industrial application. The fruit ripens quickly and has a shelf life of no more than 8 days. Therefore, it is urgently important to develop industrialized products to guarantee the surplus of fresh consumption and the use of the fruit in off-season periods (AZZOLINI; JACOMINO; BRON, 2004).

Guava is widely used for the industrialization of juices, pulps and nectars, as well as sweets, jams, jellies, fruit in syrup, purees, syrups, wines, among others (NASCIMENTO, 2010).

Brazil is one of the world's largest producers and the third largest exporter of guava, and stands out in the production of red-fleshed guavas. The main cultivars are Paluma, Sassaoka, Rica, Século XXI, Pedro Sato. Roxa, Cascão, Goiabeira do Campo and Cortibel. Since 2005, Embrapa Clima Temperado, Emater/Ascar-RS, the Eliseu Maciel School of Agronomy/Federal University of Pelotas and other institutions have been working to qualify guava cultivation in Rio Grande do Sul, especially in the municipality of Pelotas, where the Paluma, Pedro Sato and Século XXI cultivars are already found (COSTA; PACOVA, 2003).

The Paluma cultivar is the most widely grown, as it produces fruit with characteristics that allow it to be used for both industry and fresh consumption. The fruit weighs over 200 grams and is pyriform. When ripe, the skin is smooth and yellow; the flesh is deep red, firm and thick; the taste is pleasant due to the high sugar content (approximately 10°Brix), the acidity is balanced and the seeds appear in small numbers (PEREIRA et al., 2009).

2.1.1 Guava's nutritional composition

The nutritional composition of guava can vary according to a number of factors, including soil fertility, the time of year, climatic conditions, the position of the fruit on the tree, the cultivar and the stage of ripeness (SIQUEIRA, 2006). This fruit contains appreciable quantities of acids, sugars and pectins, however, the quantities of tannins, flavonoids, essential oils, sesquiterpenoid alcohols and triterpenoid acids stand out. Guava is also used in folk medicine for colic, diarrhea and dysentery. The leaves are used as an antitussive, anti-inflammatory, mouthwash and intestinal antiseptic (IHA et al., 2008). It is indicated as a potent antioxidant due to its high polyphenol content (RAMÍREZ; DELAHAYE, 2009). Studies have shown antimicrobial, antimutagenic and hypoglycemic activity, among others. It also contains high levels of vitamin A and C and appreciable amounts of B vitamins, mainly thiamine (B1), riboflavin (B2) and niacin. Minerals are a group of nutrients necessary for good health, as they play essential roles such as structural constituents of body tissues, organic regulators, muscle activity and the body's acid-base balance; components or activators/regulators of many enzymes, as well as being involved in the process of growth and body development. As components of food, minerals play a part in flavor, activate or inhibit enzymes and other reactions that influence the texture of food (AQUINO, 2008). The high levels of fiber, protein, total sugars and mineral elements such as calcium, phosphorus and potassium make this fruit one of the most complete and balanced in terms of nutritional value (MANICA et al., 1998).

2.1.2 Bioactive compounds

Fruits produce a wide variety of organic compounds through their secondary metabolism. These compounds do not seem to play a direct role in their growth and development, but they are very important in responding to predators, ultraviolet (UV) radiation, climatic adversities, among others (TAIZ; ZEIGER, 2004).

Secondary metabolites are divided into three large groups: terpenes, phenolic compounds and alkaloids (Figure 1) (LIU, 2003). Compounds of terpene and phenolic origin, as well as contributing to the color, aroma and flavour of fruit and vegetables, have the ability to stabilize the free radicals present in our bodies. Because they prevent oxidative damage to cells, they are considered important in reducing the initiation and progression of chronic disorders. From this point of view, studies have linked the consumption of fruit with a reduction in cancer, due to the presence of these compounds (NETO, 2007; VUONG et al., 2009; HE; GIUSTI, 2010).

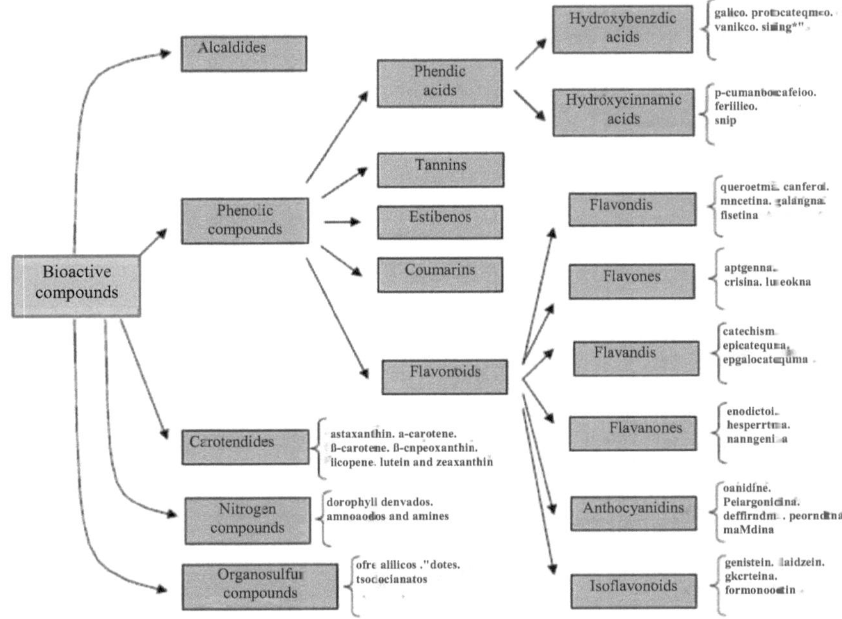

Figure 1. Classification of bioactive compounds. Source: LIU, 2003.

2.1.2. 1 Carotenoids

Carotenoids are terpenoid compounds made up of eight isoprene units. Around 600 carotenoids are found in nature, and these are classified into two groups: carotenes, which are exclusively made up of hydrocarbons; and xanthophylls, which have oxygenated functional groups in their molecule. They have hydrophobic and lipophilic characteristics and are soluble in organic solvents such as acetone, alcohol and chloroform (AMBRÓSIO; CAMPOS; FARO, 2006). The system of conjugated double bonds forms the chromophore group, responsible for the resulting color in food. A minimum of seven conjugated bonds are required for the yellow color to appear. As the number of conjugated bonds increases, the color becomes redder, consequently the absorption bands and wavelengths increase (MORAIS, 2006), and the color can vary from red to orange, yellow or brown. Because it has unsaturated compounds in its chain, the carotenoids are sensitive to light, temperature extremes, acidity and oxidation reactions (AMBRÓSIO; CAMPOS; FARO, 2006).

Despite the large number, only 40 carotenoids are found in food, and of these only 14 are bioavailable (GOMES, 2007). These micronutrients are of great importance in food, and the most commonly found are: β-carotene (carrots), lycopene (tomatoes, guava) (POPCU, 2004), xanthophylls (corn, mango, papaya, egg yolk) and bixin (annatto dye) (FONTANA et al., 2000). Figure 2 shows some of the chemical structures of carotenoids commonly found in fruit.

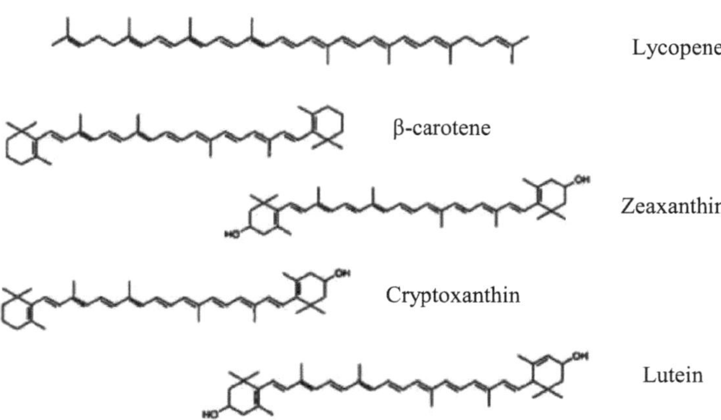

Figure 2: Chemical structure of the carotenoids most commonly found in fruit. Source: RODRIGUEZ-AMAYA; KIMURA; AMAYA-FARFAN, 2008.

Carotenoids are widely used by industries as natural colorants to replace synthetic ones, in foods and drinks, as well as cosmetics and animal feed (RIOS; ANTUNES; BIANCHI, 2009). The pink color of guava is mainly due to the carotenoids present, with lycopene accounting for around 80% of the fruit's carotenoids. The high content of this pigment in guava (54 mg.100g^{-1}) is a factor that may contribute to the increased interest in exporting this fruit (SATO; SANJINEZ; CUNHA, 2004).

Lycopene is an acyclic isomer of β-carotene, hydrocarbon, fat-soluble, melts at 172°C, has a symmetrical molecule and is optically inactive. Heat processing probably induces lycopene to isomerize to the cis form, which increases its bioavailability for absorption (MORAIS, 2006). It is cited as the carotenoid with the greatest antioxidant capacity, due to the large number of conjugated dienes, which makes it one of the greatest absorbers of *singlet* oxygen, reducing the risk of cancer, especially prostate and lung cancer. It also has the ability to reduce mutagenesis and, in high concentrations, can inhibit the growth of cancer cells (GOMES, 2007; SILVA et al., 2010), acting to prevent breast, digestive tract, cervix, bladder and skin cancers. Lycopene also plays an important role in preventing cardiovascular diseases, as well as preventing the oxidation of lipids, proteins, low-density lipoproteins (LDL) and DNA (RIOS; ANTUNES; BIANCHI, 2009).

Rodriguez-Amaya and Porcú (2004), studying the behavior of lycopene during the processing of guava, in the form of pulp, guava jam and guatchup, reported lycopene contents of these products in the order of 134, 83.9 and 98.3 µg of lycopene.g^{-1} respectively, while Jacques et al. (2009) in their study reported the presence of carotenoids in various vegetables analyzed, such as tomatoes, papaya, pitanga, mango, lettuce, carrots and guavas.

2.1.2.2 Phenolic compounds

Fruits, especially those with a red or blue color, are the most important sources of phenolic compounds in diets (DEGÁSPARI; WASZCZYNSKY, 2004). These compounds have at least one aromatic ring in which at least one hydrogen is replaced by a hydroxyl group, which gives them antioxidant power. They can be in simple or polymer forms, free or complexed with sugars and proteins. Phenolic compounds are grouped into different classes (phenolic acids, flavonoids, stilbenes and lignans) which are distinguished by the number of constituent carbon atoms together with the structure of the phenolic base skeleton (GHARRAS, 2009). The class of phenolic acids has bioactive functions, including derivatives of benzoic and cinnamic acids. Gallic acid and p-hydroxybenzoic acid are the most commonly found benzoic acid derivatives in nature. Of the cinnamic acid derivatives, four are widely distributed in plants: caffeic, ferulic, pecumaric and sinaptic acids. Flavonoids are the largest group of phenolic compounds found in plants. Variations in the substitution of the "C" ring in the structure of these compounds result in the main classes of flavonoids, i.e. flavonols, flavones, flavanones, flavonols, isoflavones and anthocyanins (MARTINS et al., 2011).

Haida et al. (2011), in their studies on white and red guava pulp, found total levels of phenolic compounds ranging from 1650 to 1740 mg gallic acid equivalent (GAA).$100g^{-1}$ of sample.

2.1.2.3 L-ascorbic acid

Vitamin C in fruit is predominantly composed of ascorbic acid (AA) and the primary product of its oxidation, dehydroascorbic acid (DHA) (Fig. 3), which also have important biological activity in oxidative stress reactions (CORDENUNSI et al., 2005). L-ascorbic acid is the main compound with 100% vitamin C activity. L-dehydroascorbic acid has the same biological activity as vitamin C, but is not very stable. After its formation, it undergoes a ring-opening reaction to form 2,3-diceto-L- gulonic acid, which has no vitamin activity.

Ascorbic acid is a water-soluble vitamin that is essential for human beings, as the human body does not synthesize it and has a low storage capacity, which is why it is necessary to take it regularly. This vitamin is essential for the various functions of the immune system, including collagen formation and healing; however, its best-known effect is its high capacity to inactivate free radicals, which can destroy cell membranes and cause lipid peroxidation (KALT, 2005; LEE; KADER, 2000).

The best sources of vitamin C are raw fruits, vegetables and legumes. Because it is very sensitive, this vitamin is easily destroyed by heat (during cooking), oxygen (air) and light (COZZOLINO, 2005). It is found mainly in free form and also bound to proteins (KALT, 2005; LEE; KADER, 2000).

Guavas are rich in vitamin C, especially when fresh, but also in the form of jams, juices, compotes and jellies, with concentrations ranging from 180 to 300 $mg.100g^{-1}$ (SOUZA, 2003).

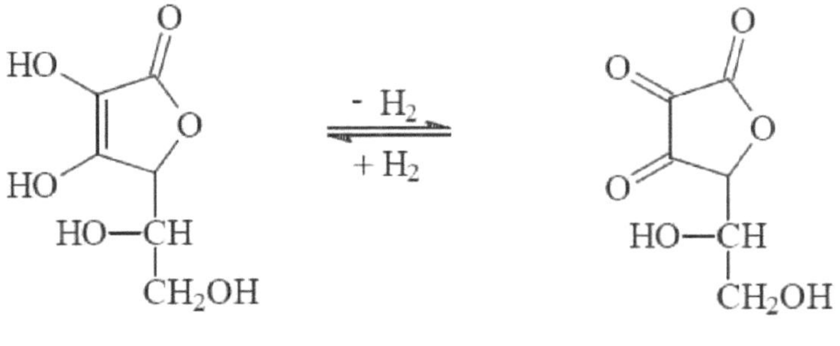

L - ascorbic acid   L - dehydroascorbic acid

Figure 3 - Ascorbic acid oxidation reaction. Source: SEVERO, 2009.

2.1.2.4 Antioxidant capacity

The antioxidant capacity of fruit and vegetables is associated with the presence of compounds that have the ability to inactivate free radicals, among which are compounds from secondary metabolism, such as phenolic compounds (phenolic acids, flavonols, anthocyanins), terpenoids (carotenes and lycopenes), vitamins (C, E and A), as well as enzymes such as superoxide dismutase, catalase and peroxidase (CORDENUNSI et al., 2005).

A diet low in antioxidant compounds, contact with pollution and smoking, among others, can generate large quantities of free radicals. These highly unstable radicals try to stabilize themselves in biomolecules, such as those present in cell membranes and even nucleic acids, and can trigger mutations, causing diseases such as cancer (DIMITRIUS, 2006).

According to studies, the greater antioxidant activity of certain fruits may be related to the total content of phenolic compounds, the content of individual phenolic compounds, the combination of these compounds or synergism with other compounds such as vitamin C (ZAICOVSKI, 2008).

Alothman et al. (2009) determined the antioxidant activity of guavas with different extracts and found that the acetonic and methanolic extracts at different concentrations obtained inhibition percentages ranging from 93 to 94 and 67 to 68% respectively.

2.1.3 Guava nectar

Fruit growing (fruit in general) is responsible for more than 505 million tons produced worldwide (FAO, 2012), with Brazil being the third largest producer (BRAZILIAN FRUIT, 2012). Allied to this is the growth of the fruit agro-industry, particularly the beverage sector. According to Rosa et al. (2006), the Brazilian domestic market for ready-to-drink juices averages around 250 million liters per year. However, according to data from the Brazilian Fruit Institute (IBRAF, 2011), in 2010, the consumption of juices of all flavors amounted to approximately 550 million liters in Brazil alone, and among the different fruit drinks, the demand for nectars has been growing (ABIR, 2011). This growth is in line with consumers who are increasingly concerned about their quality of life and opting for a healthier diet. This change in lifestyle in recent years has led the soft drink industry to invest in the production of industrialized juices and nectars (ARRUDA, 2003).

Tropical fruit nectars are generally cloudy drinks that are produced by mixing concentrated juices or pulps (ASKAR & TREPTOW, 1992). The term "fruit nectar" is used to refer to pulpy fruit juices mixed with sugar syrup and citric acid to produce a "ready-to-drink" beverage. Although these drinks are similar to fruit juices in aroma, they cannot be called such due to the presence of water, sugar and acid (LUH & EL-TINAY, 1993). Fruit juice itself is 100% fruit juice, with the exception of very viscous fruits that require dilution of their juice or pulp. When this happens, in the case of fruit of tropical origin, the drink is called tropical juice, according to Brazilian legislation (BRASIL, 2009).

Price is another factor why nectars are gaining ground among consumers, but it is precisely because they contain less pulp that their price is lower than juice. In the specific case of guava nectar, the minimum amount of pulp required by law is 35% (Brasil, 2003).

Another decisive factor for the acceptance of juices and nectars, apart from price and color, is the visual appearance of cloudy drinks, as they must not show sedimentation or phase separation. The stable distribution of particles that
cause turbidity to be an important quality criterion for the product, even if it retains its nutritional value and taste (MOLLOV & MALTSCHEV, 1996). Phase separation can occur for various reasons, as it is a complex phenomenon involving, among other factors, chemical bonds, the density of the dispersed and dispersing phases, particle size and the viscosity of the dispersing phase.

According to the Ministry of Agriculture and Supply, which regulates the activities of fruit juice and nectar processing industries, through Ordinance No. 12 of September 4, 2003, Art. 1°, Anexo II è Padrões de Identidade e Qualidade de Néctar (BRASIL, 2003), "guava nectar" is the unfermented drink obtained by dissolving guava pulp (*Psidium guajava* L.) and sugars in drinking water, intended for direct consumption, and may contain added acids", and must comply with the characteristics and composition described in Table 1.

Table 1. Identity and quality standards for guava nectar (BRASIL, 2003)

Features	
Color:	Ranging from white to reddish
Taste:	Characteristic
Aroma:	Own
Composition	**Minimum value**
Guava pulp (g/100g)	35,00
Soluble solids in °Brix at 20°C	10,00
Total acidity in citric acid (g/100g)	0,10
Total sugars (g/100g)	7,00
Ascorbic acid (mg/100g)	14,00

Source: Portaria n°12, de 04 de setembro de 2003, Art. 1°, Anexo II ě Padrões de Identidade e Qualidade ǰe Néctar.

Juices and nectars are usually preserved by adding chemical substances, by freezing or by mild thermal treatments such as pasteurization, since the microbiota capable of developing in these products has little thermal resistance (UBOLDI EIROA, 1989).

The quality of processed juices and nectars is attributed to their physical and chemical parameters, such as acidity, pH, soluble solids, sugars, color, viscosity and ascorbic acid content, as well as microbiological and sensory characteristics (EWAIDAH, 1992 cited by MORI, 1995). Prior knowledge of the content of ascorbic acid in processed foods is very important, because as well as playing a fundamental role in human nutrition, its degradation can promote non-enzymatic browning. Ascorbic acid is also an important indicator of quality, as it is the most thermolabile vitamin and its presence in the food indicates that the other nutrients are probably also being preserved (ALLAH & ZAKI 1974, BENDER 1978, GUTHRIE 1989, cited by CARDELLO & CARDELLO 1998).

2.1.4 Hydrocolloids

Hydrocolloids, also known as gums, mucilages and hydrophilic polysaccharides, have been widely used in the food industry because, among other functions, they provide a gel structure, increase viscosity, act as an encapsulating agent, enable film formation, control crystallization, inhibit syneresis and increase physical stability (DZIEZAK, 1991; DICKINSON, 2003). These polymers have the capacity to directly influence food properties such as appearance and texture.

Generally, hydrocolloids are used in products with partial or total fat reduction, in order to minimize changes to the texture of the product and avoid phase separation in emulsions (KATZBAUER, 1998; TONELI et al., 2005). Positive effects are also obtained in foods that are consumed frozen or that are preserved by freezing in order to reduce the negative effects related to phase transition (HERCEG et al., 2000). However, the influence of these hydrocolloids can depend on the interactions between them and also between other biopolymers and/or other food components. For this reason, the selection of hydrocolloids for a specific application involves more than just the selection of functional properties. Among the factors that may be most relevant and should be taken into account in this choice are the appearance of the final product, the type of product application, optical properties of the final product, viscosity, taste, texture, odor, emulsifying properties, compatibility with the system, stability under different storage conditions, use as a preservative, toxicity of the product and cost (TONELI et al., 2005).

Hydrocolloids fall into three categories: natural (plant exudates: arabica, gatti, karaia, tragacanth and larch; algae: agar-agar, carrageenan, alginates and furcelaram; plant extracts: locust, guar and pectin; cereals; tubers; roots: starch), modified (cellulose derivatives: carboxymethyl cellulose, hydroxyethyl cellulose; starch derivatives: hydroxyethyl starch and amylose xanthate; animal by-products: water-soluble chitin derivatives, gelatin) and synthetic (microbial products: xanthan, dextran, curdlan, gelan) (GLICKSMAN, 1982, cited by GODOY, 1997).

In fruit processing, the use of hydrocolloids is already well used to improve the texture of pulp and juice concentrate (HUI et al., 2006). The advantage of using these agents is that even in small quantities they can keep the pulp suspended without significantly altering the quality of the product.

2.1.4.1 Xanthan gum

Xanthan gum is an extracellular polysaccharide produced by microorganisms of the *Xanthomonas* species, and is commercially produced by *Xanthomonas campestris* (RIBEIRO; SERAVALLI, 2004). It is a gum that is soluble in both cold and hot water, its solutions are highly pseudoplastic and its viscosity is very stable over wide pH and temperature ranges (SWORN, 2000).

This gum consists of a heteropolysaccharide whose primary structure is a chain made up of repeated β-D-glucose units linked by β 1-4 bonds, with branches made up of β-D-mannose ě 1,4-β-D-glycuronic acid ě 1,2-α-D-mannose, and may contain pyruvic and acetic acid. Its branches occur in alternating β-D-glucose units, and the bonds with the β-D-mannose units are of the 0-3 type (Fig. 4). Due to the presence of carboxylic groups in its structure, xanthan gum is considered an anionic polymer (GARCÍA-OCHOA et al., 2000).

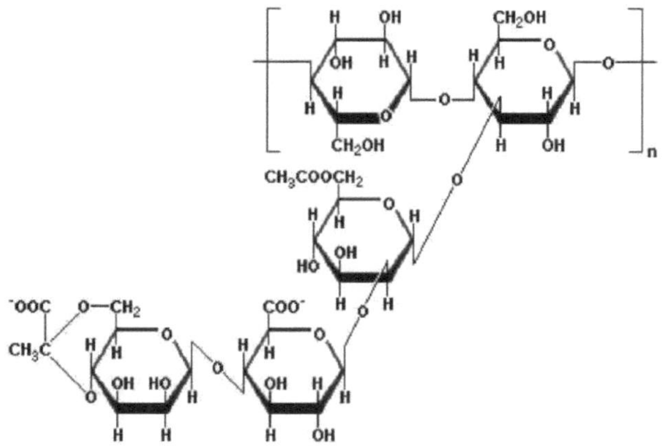

Figure 4 - Chemical structure of the xanthan molecule. Source: NERI et al., 2008.

Xanthan gum is used as a thickening and stabilizing agent in foods, and is also used in pharmaceutical formulations, cosmetics and agricultural products (GARCÍA-OCHOA et al., 2000). Godoy et al. (1998) studied the stabilization of guava nectar with xanthan gum (0.075%, 0.125% and 0.175%); waxy starch (0.75%, 1.25% and 1.75%) and carrageenan (0.125%, 0.175% and 0.225%), and found that xanthan gum at a concentration of 0.175% stabilized 99% of the volume of the nectar, as well as guaranteeing preference in the sensory test.

2.1.4.2 Guar gum

Guar gum (Fig. 5) extracted from the grains of the guar plant (*Cyamposis tetragonolobus*) is a galactomannan made up of linear chains of D-mannopyranosyl units linked by β 1 -4 bonds and D-galactopyranosyl units linked by α 1 -6 bonds. It is a linear galactomannan made up of D-mannose and D-galactose units in a 2:1 ratio.

It is a high molecular weight gum, heat stable, capable of forming colloidal dispersions in water with high viscosity, it does not form gels and the viscosity is little affected by pH at the extremes between 4-9. The interaction of this gum with xanthan produces more viscous solutions with high pseudoplasticity, which is important for the production of foods with different textures, as well as the reduction ofnfood product costs by adding it within the limits allowed by legislation, which is 0.1% (DEA; MORRISON, 1975; ROCKS, 1971). The ability to hydrate quickly in cold aqueous systems, generating highly viscous solutions, is the most important property of this gum (GOLDSTEIN, ALTER and SEAMAN, 1973).

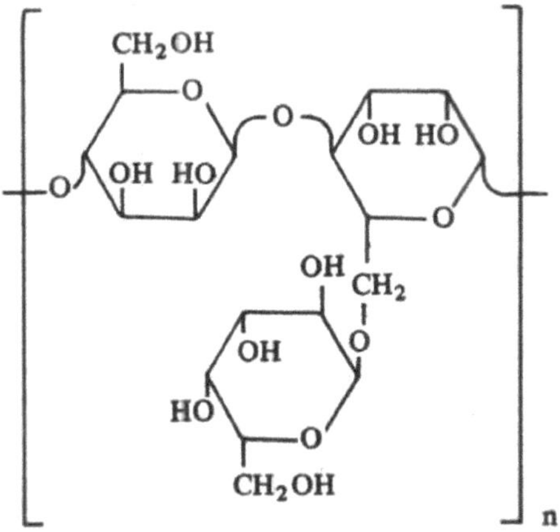

Figure 5 - Chemical structure of guar gum.
Source: RIBEIRO & SERAVALLI, 2004.

Guar gum has been used as a thickener and stabilizer in drinks, sauces and ice cream (BOBBIO and BOBBIO, 1992).

2.1.4.3 Pre-gelatinized rice flour

During the processing of rice grains, by-products are generated that can be utilized, and among these are the broken grains (14 to 60%) that can be used to produce rice flour (PESTANA; MENDONÇA; ZAMBIAZI, 2008).

Different technologies have been used to obtain pregelatinized flour, including thermal processing, which modifies physical, chemical and sensory parameters, giving these flours different characteristics; extrusion stands out among them. During this process, changes occur in the structure of the food, altering the conformation of proteins, interactions between carbohydrates, inactivation of enzymes and development of new texture characteristics (TAVARES, 2010).

According to Tavares (2010), one use of these raw and modified rice flours is in instant products, which have become very popular recently, including ready-made mixes for puddings, flans and others that do not require the technological properties of gluten (DORS et al., 2006). However, there is no reference to its use in the stabilization of fruit nectars.

2.1.5 Plant cell wall and pectinolytic enzymes

The cell wall is responsible for the strength and rigidity of plant tissues, as well as supporting the structure of the cells, performing important functions such as absorption and transportation of water and minerals, secretions, enzymatic activity and others (CHITARRA & CHITARRA, 2005). As can be seen in figure 6, the cell wall is made up of several layers and different macromolecules, and is responsible for the fruit's textural characteristics. The middle lamella is the outermost layer that provides cohesion between the cells and is therefore known as cell cement, being composed mainly of pectins (CHITARRA & CHITARRA, 2005).

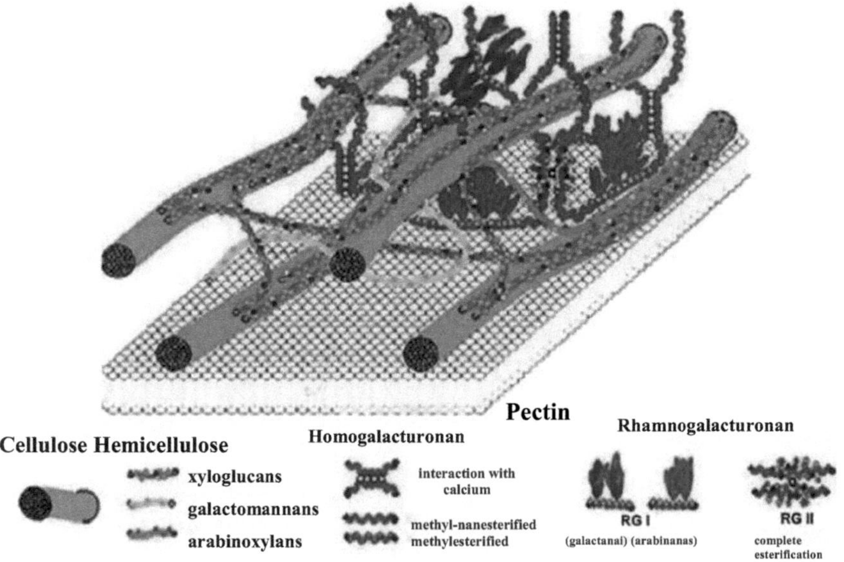

Figure 6- Structure of the plant cell wall. Source: BARBOSA, 2010.

Chemically, pectic substances are colloidal complexes of acidic polysaccharides, composed of galacturonic acid residues linked by α-1,4 bonds, partially esterified by methy ester groups (GUMMADI & PANDA, 2005) and partially or completely neutralized by one cr more bases (KASHYAP et al., 2001). The American Chemical Society has classified pectic substances into: protopectin, pectinic acid, pectic acid and pectin, all three of which are totally or partially soluble in water (KASHYAP et al., 2001; ALKORTA et al., 1998). The general term pectin (Figure 7) refers to water-soluble pectinic acids with a variable degree of methyl ester groups and a degree of neutralization capable of forming gels with sugars and acids under suitable conditions (SAKAY et al., 1993). It consists of a structure of axial bonds of α-1,4-D-galacturonic acid units and contains L-rhamnose, arabinose, galactose and xylose molecules as side chains (GUMMADI, 2005).

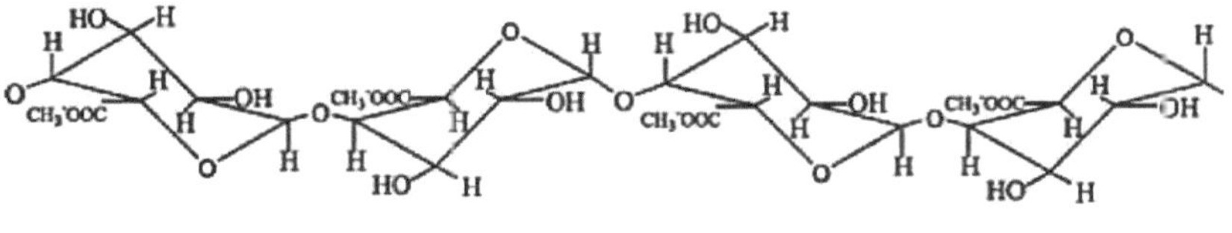

Figure 7. Structure of the pectin molecule Source: UENOJO E PASTORE, 2007.

Pectic substances are responsible for the consistency, turbidity and appearance of fruit juices and their presence causes a considerable increase in juice viscosity, making filtration and concentration difficult. The addition of pectinolytic enzymes to fruit and vegetable purees results in the degradation of pectin present in the middle lamella and primary cell wall of higher plants and other high molecular weight components, decreasing viscosity and increasing juice yield, causing a crystalline appearance in the final product and reducing filtration time by up to 50% (JAYANI; SAXENA; GUPTA, 2005; WILLATS et al., 2001). During the mid-1930s, when the fruit industries began producing juices, yields were low and many difficulties were encountered in filtering the juice and achieving stable clarification. Since then, research using pectinases, cellulases and hemicellulases from micro-organisms, together with knowledge of the plant components of fruit, has reduced these difficulties without increasing costs, thus reducing processing time and improving the extraction of fruit components (BHAT, 2000; JAYANI et al., 2005; HEERD et al., 2012). Other applications of these enzymes in the food industry include fruit ripening, tomato pulp extraction, tea and chocolate fermentation, vegetable waste treatment, vegetable oil extraction and baby food production. They can also be used in the fermentation industry, for example in the fermentation of cocoa, coffee and tobacco, as well as in the degumming of natural fibers in the textile and paper industry (USTOK; TARI; GOGUS, 2007).

Pectinases are a group of enzymes that degrade pectic substances by hydrolyzing the glycosidic bonds along the carbon chain. They can be depolymerizing or deesterifying and are produced by plants, filamentous fungi, bacteria and yeasts (UENOJO & PASTORE, 2007). Fungi Filamentous organisms, especially *Aspergillus niger,* are the main producers of pectinases used in the fruit juice and wine industries (KASHYAP et al., 2001). Jayani, Saxena and Gupta (2005) state that 25% of all enzymes marketed are pectinases of microbial origin. Among the companies making these enzyme preparations are those based in Europe, the United States and Japan (UENOJO; PASTORE, 2007).

According to the literature, fungal pectinases are stable under acidic pH conditions (GUMANDI; KUMAR, 2006) and are susceptible to denaturation at temperatures above 50°C (GALIOTOU-PANAYOTOU; KAPANTAI; KALANTZI, 1997). Jayani, Saxena and Gupta (2005) mention that most enzymes have an optimum pH range of 3.3-5.5 and an optimum temperature range of 30-50°C.

Pectinases include various enzymes divided into classes and subclasses depending on substrate specificity and mode of action, such as methylesterases, hydrolases and lyases (YADAV et al., 2009). Pectinamethylesterases break the ester bond by demethoxylating galacturonic acids esterified with methanol. They result in the formation of pectins with a low content of methoxylation or pectic acid, as well as methanol. Polygalacturonases and pectatolyases preferentially attack the bonds between galacturonic acids in pectins with a low methoxylation content (or pectic acid), while pectinolyases prefer bonds between galacturonic acids in methoxylated pectins (pectins with a high methoxylation content) (KOBLITZ, 2010). Pectinolyase is the only enzyme capable of hydrolyzing highly esterified pectins without the prior action of another pectinase (BRAVO et al., 2000). The mode of action of these enzymes is illustrated in Figure 8.

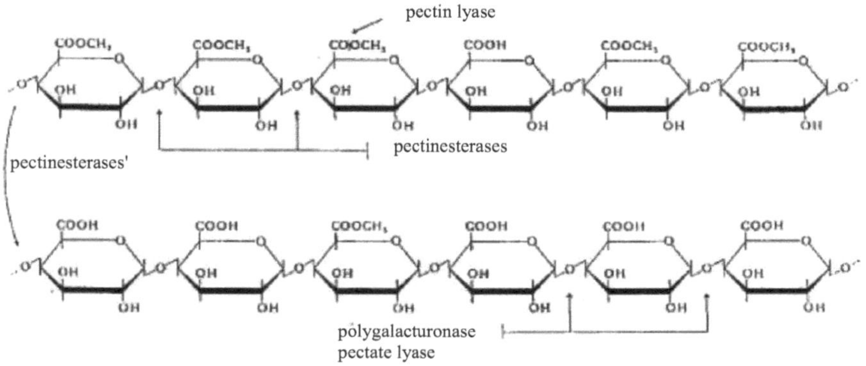

Figure 8. Attack point of pectinases on the pectin molecule. Source: MARTINS, 2006.

In a study of the stability of cajá juice treated enzymatically with 120 ppm of a pectinolytic enzyme complex (Pectinex Ultra SP-L) for 30 min. at 25°C, Silva et al. (1997) observed that after 120 days of storage, the physical-chemical and sensory characteristics of the juice remained without significant alterations, indicating good stability. In plum, banana and guava juices investigated by Amin and Essa (2002), the effects of pectinase (Clarex ML) improved various quality parameters of these juices.

Chapter 3

3 Material and methods

The experiment in this study was carried out at Embrapa Clima Temperado, at the base of the Cascata Experimental Station, at the Federal Institute of Education, Science and Technology Sul-Rio-Grandense, Pelotas-Visconde de Graça campus (CaVG), and the analyses were determined at the Chromatography Laboratory of the Department of Agroindustrial Science and Technology (DCTA) of the FAEM (Eliseu Maciel Agronomy Faculty) of the Federal University of Pelotas, Capão do Leão Campus.

3.1 Material

The guavas of the Paluma cultivar were obtained from a rural property in the municipality of Pelotas/RS-Brazil (latitude: 31°29'12"S, longitude: 52°32'57"O and altitude: 226m), during the 2013 and 2014 harvests.

The stabilizing agents used were xanthan (Sigma Aldrich), guar (Sigma Aldrich), pregelatinized rice flour and a complex of pectinolytic enzymes, consisting mainly of pectinolyase, polygalacturonase and pectinamethylesterase and small amounts of hemicellulases and cellulases (Sigma Aldrich). To determine the antioxidant capacity, 2,2-diphenyl-1- picrylhydrazyl (DPPH) (Sigma Aldrich) was used as the radical. The other reagents used to carry out the spectrophotometer analyses were PA grade.

3.2 Methods

The guavas, weighing around 20 kg, were harvested by hand in the early hours of the day (from 7 to 8 am), when the temperatures are milder, packed in PVC boxes and transported to the processing unit belonging to the Federal Institute of Education, Science and Technology of Rio Grande do Sul, Pelotas-Visconde de Graça campus (CaVG), Pelotas-RS. Here, they were manually selected for health, physical integrity, uniformity of color and degree of ripeness. They were then washed with drinking water to remove surface dirt and then immersed in a solution of chlorinated water with 500 ppm of active chlorine for 10 minutes. They were then rinsed with drinking water to remove the remaining chlorine. After this stage, the fruit went through a pulper with a 0.8 mm mesh, where the guava pulp was obtained, which was packed in polyethylene bags with a capacity of 1 Kg and stored in a freezer at -18°C until the time of analysis and preparation of the nectars.

The nectar was processed (Fig. 10) on the premises of the mini factory at the Cascata Experimental Station, Pelotas-RS. Table 2 shows the different formulations (treatments) applied to the nectars. The base formulation (control) consisted of 55% water, 35% guava pulp and 10% sugar (sucrose). Stabilizers (xanthan, guar, pregelatinized rice flour) and/or the enzyme pectinase were added to the other formulations. The gums were dispersed in the sugar and slowly dissolved in the guava pulp already added to the water to prevent lumps from forming. They were then homogenized using an industrial blender and heat-treated in a bain marie (90°C/10 min.), followed by hot filling (85°C) into previously sterilized 150 mL glass bottles and closed immediately with screw-on metal caps. After closing, the bottles were cooled by immersion in water at 45 and 25°C respectively, according to the methodology described by Sousa et al. (2006). For the formulations containing the pectinase enzyme, 1400 ppm of the enzyme (pectinase) was added, along with the water and sugar, before pasteurization, remaining in a water bath at 35°C/2 hours, according to the methodology described by Rodrigues (2013), with some adaptations and then following the same procedures described above for the nectar formulations containing the gums.

After the nectars had been prepared for 24 hours, the glass bottles were transported to the Chromatography Laboratory of the Department of Agroindustrial Science and Technology at the Federal University of Pelotas, Pelotas-RS. The drinks were shaken for 3 minutes and stored in a room with clear glass windows, in which natural light as well as artificial light was shining in, at an ambient temperature of 22°C ± 3.6°C, which was measured daily using a temperature meter (Datalloger) for a period of up to 180 days, in order to simulate the environment in which nectars are normally exposed for consumption in some sales outlets.

Table 2 - Guava nectar formulation: gum and enzyme concentrates

Formulations *	Quantity
T1- Control	-
T2- Xanthan	0,1%
T3- Guar	0,1%
T4- Rice Flour	0,1%
T5- Xanthan + Guar	0,05 + 0,05 %
T6- Xanthan + Rice Flour	0,05 + 0,05 %
T7- Guar + Rice Flour	0,05 + 0,05 %
T8- Enzyme	1400 ppm
T9- Enzyme + Xanthan	1400 ppm + 0.1 %
T10- Enzyme + Guar	1400 ppm + 0.1 %
T11- Enzyme + Rice Flour	1400 ppm + 0.1 %

* The nectars were formulated in accordance with Ordinance No. 12 of September 4, 2003, Art. 1 Annex II - Standards of Identity and Quality of Nectars.

The simplified processing flowchart for guava pulp and nectar is shown in Figures 9 and 10, respectively.

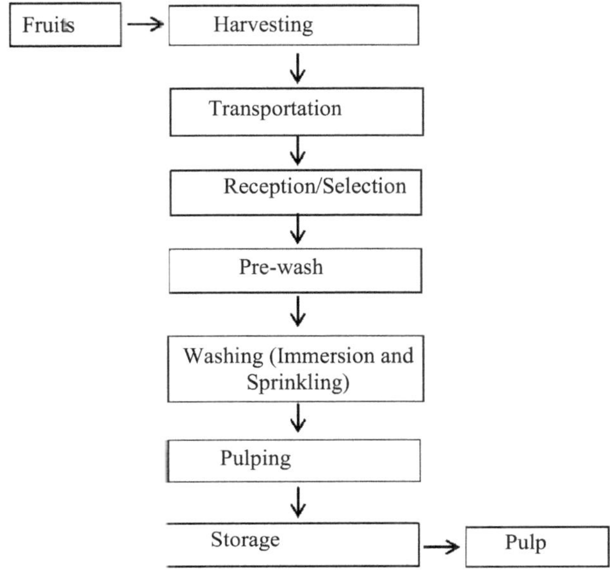

Figure 9. Flowchart for processing guava pulp (*Psidium guajava* L.) cv. Paluma

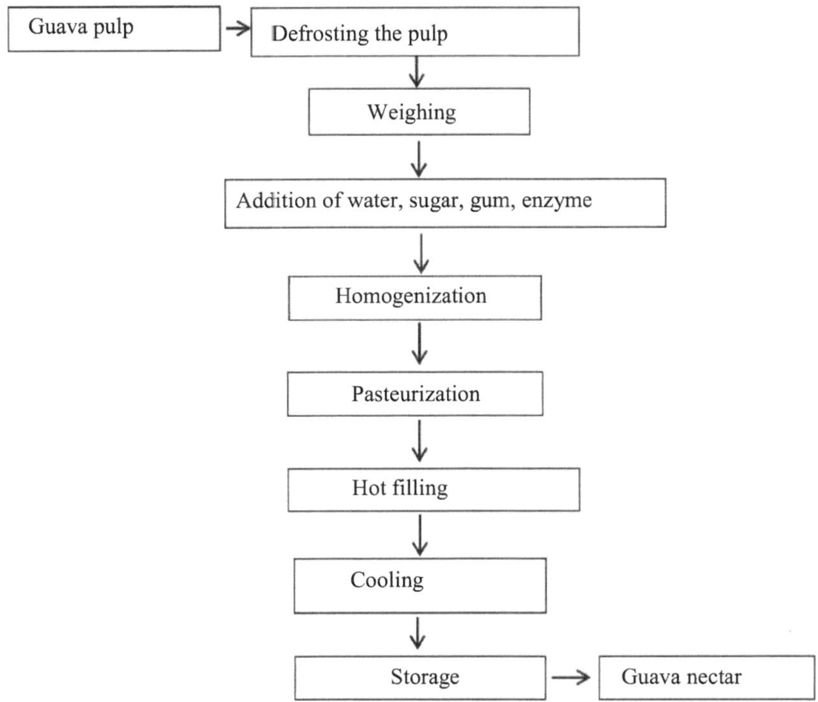

Figure 10. Flowchart for processing guava nectar (*Psidium guajava* L.) c.v Paluma.

3.2.1 Physico-chemical analysis

Analyses were carried out on the pulp of the Paluma guava cultivar and then on the guava nectars at 45-day intervals over a period of 180 days, with the initial time being analyzed 24 hours after processing. The analyses were carried out in triplicate.

3.2.2 Hydrogen potential (pH) and total titratable acidity

To check the hydrogenic potential, five grams of sample were homogenized in 25 mL of distilled water and read in a potentiometer (PHMETERDIGIMED DM-20), calibrated with 4.0 and 7.0 buffers at a temperature of 20° C. The homogenate obtained above was used to determine the total titratable acidity, using NaOH 0.1N to pH 8.1. The total titratable acidity results were expressed in mg of citric acid per 100g of sample on a wet basis (INSTITUTO ADOLFO LUTZ, 2008).

3.2.3 Total soluble solids (TSS)

The soluble solids content was determined by refractometry using an Atago Palette digital refractometer, PR-32 α. After calibrating the refractometer with distilled water, the sample was poured over the surface of the prism and the reading expressed in °Brix.

3.2.4 Color

Color was assessed by instrumental method using a colorimeter (Minolta® CR - 300), directly obtaining the coordinates of the CIE L*a*b* space, where the a* and b* coordinates vary, respectively, from (-) green to (+) red and from (-) blue to (+) yellow, and L* (lightness index) varies from black (0) to white (100). The a* and b* values were used to calculate the Hue angle (°h* = tang b*. a*). Chroma (C*) expresses the saturation or intensity of the color, while the hue angle (h°) indicates the observable color and is defined as starting on the +a* axis, in degrees, where 0° is +a* (red), 90° is +b* (yellow), 180° is -a* (green), and 270° is -b* (blue).

3.2.5 Phenolic compound content

The methodology described by Swains and Hillis (1959), with a few modifications, was used to quantify the total phenolic compounds. For the extraction of phenolic compounds, five grams of sample were added to 15 mL of methyl alcohol and homogenized for one minute in an Ultra Turrax® (IKA®, T18 digital), then centrifuged for 20 minutes at 7830 rpm in a centrifuge (Eppendorf, 5430) cooled to 21°C. For the reaction, a 250 µL aliquot of the supernatant was diluted in 4 mL of distilled water and the control prepared with 250 µL of methanol. 250 µL of 0.25 N Folin-Ciocalteau was then added and homogenized in a vortex (Phoenix, AP-56), after 3 minutes of reaction 500 µL of 1.0 N sodium carbonate was added. After two hours of reaction at room temperature, the absorbance was read on a spectrophotometer (JENWAY, 6700 UV/Vis) at a wavelength of 725 nm. A standard curve prepared with gallic acid was used to quantify the phenolic compounds, and the results were expressed in mg of gallic acid equivalent per 100g of sample on a wet basis.

3.2.6 Carotenoid content

The total carotenoid content was analyzed according to the AOAC method (970.64). Three grams of sample were homogenized with 15 mL of extracting solvent (hexane:ethanol:acetone:toluene, in the ratio 10:6:7:7), and then 1 mL of KOH in 10% (w/v) methanol was added and homogenized in a vortex (Phoenix, AP- 56) for one minute The mixture was then heated in a water bath (Nova ética, 500/1D) at 56°C for 20 minutes. After removal from the water bath, the mixture remained at room temperature for 1 hour. Aliquots of 15 mL of hexane were added to the flasks and the flasks were equilibrated to 50 mL with 10% (w/v) Na_2SO_4 in water. The flasks were homogenized and kept in the dark for 1 hour. An aliquot of the supernatant was evaluated in a spectrophotometer (JENWAY, 6700 UV/Vis) at a wavelength of 470 nm. A standard curve prepared with lycopene was used to quantify the carotenoids, and the results were expressed as equivalent mg of lycopene per 100g of sample on a wet basis.

3.2.7 Antioxidant capacity

The antioxidant capacity was determined through the ability of the compounds present in the samples to scavenge the stable radical DPPH- (2,2-diphenyl-1-picrylhydrazyl), according to the method described by Brand-Williams, Cuvelier and Berser (1995). For the reaction, 100 μL of the same extract used to evaluate the phenolic compounds was added to 3.9 mL of DPPH solution in methanol (100 mM). The solution was homogenized and the vials kept in the dark. The absorbance was measured in a spectrophotometer (JENWAY, 6700 UV/Vis) at a wavelength of 517 nm after 24 hours of reaction. The results were calculated according to equation 1 and expressed as a percentage of DPPH radical inhibition.

% DPPH inhibition = [(A0 ě A1) / A0 x 100] (eq. 1)

Where: A0 = absorbance of the control; A1 = absorbance of the sample.

3.2.8 L ascorbic acid

Quantification of ascorbic acid (vitamin C) was carried out using the Lorenz-Steves titrimetric method (ZAMBIAZI, 2010), based on the reducing action of ascorbic acid, using a standard solution of iodine and sodium thiosulfate and starch solution as an indicator. The results were determined using equation 2 and expressed in mg of L-ascorbic acid per 100g of sample on a wet basis.

mg ascorbic acid $100mL^{-1}$ juice (X) = [Y x 0.88 mg] / mL (eq. 2)

Where: Y= (total volume of iodine solution x solution factor) ě (volume of thiosulphate x solution factor)
Each mL of 0.01 N iodine solution corresponds to 0.88 mg of ascorbic acid.

3.2.9 Mineral salts

Mineral salts were determined using samples that had been previously freeze-dried, macerated and homogenized. The analytical procedure and quantification of the samples were carried out on the premises of the Embrapa Clima Temperado analytical center, Pelotas-RS. The digestion of the samples for the quantification of microelements (Cu, Fe, Mn, Zn) was carried out with nitric acid and perchloric acid and the digestion for the determination of macronutrients (Ca, Mg, K, P) was carried out with hydrogen peroxide and sulfuric acid (TEDESCO et al, 1995). The digestion of the samples was carried out in blocks and the quantification of the elements was determined by Atomic Absorption Spectroscopy (Varian, AA240FS), with the exception of potassium, which was determined by atomic emission.

3.2.10 Sedimentation

The clarified phase of the nectars was measured approximately twice a week for a period of 90 days, which was enough time for the nectars to stabilize. In order to facilitate discussion of these evaluations, the results have been expressed as a percentage of the nectar's stabilized phase (not clarified phase).

3.2.11 Viscosity

The viscosity of the nectars was assessed in a rotational rheometer (Haake® RS150), using a plate-to-plate system, PP35TI sensor, at 25°C. Viscosity was determined in the rotating module at a strain rate of 0.01 to $100s^{-1}$, for 300s, totaling 100 acquisition points. Determinations were made on the fluid portion of the sample after centrifugation at 45 and 135 days of storage.

3.2.12 Statistical analysis

The results were expressed as means and standard deviations for the determinations carried out in triplicate. The results were subjected to the Tukey and Dunnett mean comparison test with a significance level of 5%, using the SAS v8 statistical program.

Chapter 4

4 Results and discussion

4.1 Physico-chemical parameters, content of bioactive compounds and mineral salts in the raw material

The results obtained, with their respective standard deviations for physical-chemical parameters, the content of biotic compounds and mineral salts in the guava pulp cf the Paluma cultivar used for processing the nectars, are shown in Table 3 and 4.

Table 3. Physicochemical parameters of guava pulp cv. Paluma

Characteristics Evaluated	* Mean ± SD
pH	3,66 ± 0,01
Total Titratable Acidity (% citric acid)	0,40 ± 0,00
Soluble Solids (°Brix)	9,25 ± 0,05
TSS/ATT ratio	23,32 ± 0,24
Color L*	47,11 ± 0,40
Color a*	22,20 ± 0,54
Color b*	9,21 ± 0,15
°Hue	22,53 ± 0,17
Chrome	24,04 ± 0,55

*Mean values obtained from the analysis of 3 repetitions (n=3); SD = standard deviation; TSS =total soluble solids; ATT = total titratable acidity.

Paluma guava pulp had a pH (3.66) and total titratable acidity (0.40 % citric acid) slightly lower than those found in the literature, which reports pH levels ranging from 3.72 to 4.20 (MACHADO et al., 2007) and 0.48% citric acid for total titratable acidity (OSORIO; FCRERO; CARRIAZZO, 2011). Jacobo; Ahumada; Maldonado (2009) also reported total titratable acidity values of 0.38 to 0.90g.100g^{-1} for various guava selections, the cultivar of which was not mentioned. According to Bialves et al. (2012), the acidity content of guava depends very much on the stage of ripeness of the fruit, because as it ripens, the total acidity tends to reduce, since organic acids are one of the main extracts used in the respiratory process of the fruit.

According to Silva; Magalhães; Gonçalves (2009) a pH between 3.30 and 3.50 is desirable for industrial processing, so the pulp used in this study had a pH value just above the optimum processing range. Pereira et al. (2006) also found pH values above the desired level for frozen guava pulp, 3.93. For fresh fruit, Lima; Assis; Gonzaga Neto (2002) reported values of 3.88 for the Paluma cultivar and Rojas-Barquera & Narváez-Cuenca (2009) a value of 3.60, close to that found in this study. As well as defining the thermal processing of a product, pH also has a direct impact on its characteristics, such as viscosity and stabilization.

Total soluble solids (TSS) are water-soluble substances found in fruit, such as sugars (the majority compound), vitamins, acids, amino acids and some pectins (PRADO, 2009), and this content is used in the agro-industry to intensify quality control of the final product, process control, ingredients and others (CHAVES et al., 2004).

The average value of 9.25 °Brix for total soluble solids obtained in this study is within the recommended range for fruit destined for juice processing, since according to Lima et al. (2002), fruit destined for this technological purpose should have total soluble solids values higher than 8 °Brix. Serrano et al. (2007), working under different cultivation systems, times and intensity of fruiting pruning of the Paluma guava tree, obtained an average of 12.72 °Brix, while Cavalini (2004), Osório, Forero, Carriazzo (2011), Bialves et al. (2012) and Silva et al. (2009), working with the same guava variety, obtained averages in the 7.15 to 10.40 °Brix range.

The TSS/TTA ratio provides an assessment of fruit flavor and is more representative than measuring sugar content and acidity alone. This ratio expresses the balance between the content of total soluble solids and total titratable acidity (CHITARRA and CHITARRA, 1990). An average TSS/TTA ratio of 23.13 was observed in the guava pulp, exceeding the average reported by Pereira et al. (2003), who found soluble solids contents close to 10°Brix and total titratable acidity of 0.474g of citric acid/100g of pulp, resulting in a TSS/TTA ratio close to 20.0 (18.6).

The pulp showed L* values of 47.11, a* of 22.10 and b* of 9.21. The luminosity parameter (L*) indicates how close a sample is to white (100) or black (0). The chromaticity parameter a* represents positive values for red and negative values for green. The b* chromaticity parameter represents positive values for yellow and negative values for blue. Analysis of the L* parameter showed that the red guava pulp tended slightly towards black, while the a* and b* parameters tended towards red and yellow, respectively.

For the luminosity parameter, Osorio; Forero and Carriazo (2011), Tasca (2007) and Pereira et al. (2006) reported values of 38.20 to 59.41 for Paluma guava pulp. For the chromaticity parameter a*, Tasca (2007) found values of 20.80 to 31.90 for fresh Paluma guava pulp. Freda (2014) reports b* chromaticity parameters of 7.98, while Osorio; Forero; Carriazo (2011) report a much higher value than this study, 21.31.

Paluma guava pulp showed a value of 22.53 for the °Hue variable, which expresses the intensity of the hue or the color itself, justifying the red color of the guava pulp. By definition, starting on the +a axis, in degrees, 0° (+a) corresponds to red, 90° (+b) corresponds to yellow, 180° (a) corresponds to green, and 270° (-b) corresponds to blue (BIBLE; SINGHA, 1997). A value of 24.04 was found for chroma, which expresses the relationship between the values of a* and b*, thus obtaining the real color of the object analyzed, i.e. the intensity of the color or saturation.

Table 4. Characterization of bioactive compounds and mineral salts in guava cv. Paluma pulp

Characteristics Evaluated	*Mean ± SD
L-Ascorbic Acid (mg.100g)$^{-1}$	206,91± 22,09
Phenolic compounds (mg.100g)$^{-1}$	135,93 ± 6,52
Carotenoids (µg.g)$^{-1}$	102,71 ± 13,30
Antioxidant activity (% inhibition)	28,68 ± 4,13
Calcium (mg.100g)$^{-1}$	87,00 ± 0,01
Magnesium (mg.100g)$^{-1}$	91,00 ± 0,027
Potassium (mg.100g)$^{-1}$	1943 ± 0,97
Phosphorus (mg.100g)$^{-1}$	429,00 ± 0,50
Copper (mg.100g)$^{-1}$	0,361 ± 0,16
Manganese (mg.100g)$^{-1}$	1,63 ± 0,72
Zinc (mg.100g)$^{-1}$	2,72 ± 0,35

*Average values obtained from the analysis of 3 repetitions; SD = standard deviation.

The pulp of guava cv. Paluma had a high ascorbic acid content (206.91 mg L-ascorbic acid.100g^{-1}), which places it among the fruits considered to be rich in vitamin C. Oliveira et al. (2011), analyzing the same cultivar. Mariano et al. (2011) and Kuskoski et al. (2006) reported lower values than this study, between 64.0 and 83.0 mg.100g^{-1} . Fiorucci (2003) reports values close to those found in this study (218 mg L-ascorbic acid.100g^{-1}). However, McCook-Russell (2012) reports a much higher content of 1200 mg L-ascorbic acid.100g^{-1} . However, the content varies depending on soil and climate conditions, harvest time, post-harvest handling and the stage of ripeness of the fruit, since the riper the fruit, the lower the vitamin C content, due to the oxidation of these compounds during the ripening of the fruit (AZZOLINI; JACOMINO; BRON, 2004) When comparing the vitamin C content found in guava with other studies, it can be seen that it is higher than the content of fruits that have been identified as rich sources of vitamin C, such as oranges (62.5 to 85 mg L-ascorbic acid.100g^{-1}), mandarin (21.5 to 32.5 mg L-ascorbic acid..00g^{-1}) and strawberry (30 mg L-ascorbic acid.100g^{-1}) reported in the respective studies, Canniatti-Brazaca, 2010; Valente et al., 2013; Cardoso et al., 2011.

Other important compounds in the human diet are phenolic compounds, which have been cited as being largely responsible for the antioxidant activity of vegetables and fruit (FERNANDES, 2007). The content of phenolic compounds determined in this study (135.93 mg gallic acid.100g^{-1}) was lower than that found in several other studies with different guava cultivars, such as McCook-Russell et al., 2012 (195 mg GAE.100g^{-1}), Thaipong et al., 2006 (170 to 345 mg GAE.100g^{-1}), Prado, 2009 (560 mg GAE.100g^{-1}) and Freire et al., 2012 (848 mg GAE.100g^{-1}). There is a wide variation in the content of these compounds in guavas, which can also vary according to various factors such as ripeness stages, growing conditions and location, cultivars, climatic conditions, analysis techniques, among others (OLIVEIRA et al., 2011). However, studies show that guava has a higher content of phenolic compounds when compared to other fruits, such as: pineapple (362 mg GAE.100g^{-1}), mango (557 mg GAE.100g^{-1}), passion fruit (367 mg GAE.100g^{-1}), melon (126 mg GAE.100g^{-1}) and showing lower content than pitanga (115 mg GAE.100g^{-1}) and acerola (1560 mg GAE.100g^{-1}) (PRADO, 2009; OLIVEIRA et al., 2011), most of these results did not corroborate those of the respective study. There are also studies which mention that the skin of guava has a higher polyphenol content than the pulp (TASCA, 2007), and that the leaves also contain total polyphenols (HAIDA et al., 2011); both parts were not analyzed in this study.

The guava pulp had a higher carotenoid content (102.71 µg.g^{-1}) than that found by Tasca, (2007) who detected (86.60 µg lycopene.g^{-1}) and by Porcu; Rodriguez-Amaya, 2004 (61.00 and 73.00 µg.g^{-1}), both for the Paluma cultivar. However, these contents were higher than those reported for fresh guavas (16.00 µg.g^{-1}) by Pereira (2009). Silva and Naves (2001) found a higher carotenoid content for red guava (62.1 µg.g^{-1}) than some other fruits such as pitanga (16.4 µg.g^{-1}), mango (19.1 to 26.3 µg.g^{-1}) and papaya (8.5 µg.g)$^{-1}$

Fruits such as guava have various types of antioxidant compounds in their composition, including vitamin C, carotenoids and phenolic compounds. Therefore, the choice of extraction method for these bioactive compounds is of the utmost importance in order to determine the total antioxidant capacity of the fruit (MELO et al., 2008). When analyzing the antioxidant capacity using the DPPH method, it was observed that the guava pulp had a low capacity to sequester this radical (28.68 %), which could be partially related to the fact that the extract used in this study was alcoholic and did not include lipophilic compounds, such as carotenoids. However, the low percentage of inhibition found does not corroborate other studies with guavas which also used the DPPH method, such as Prado, 2009 (84%) and Melo et al., 2008 (90%).

Taking into account the importance of minerals, these were determined in the guava pulp and high levels of the following minerals were found: calcium (87 ± 0.01), magnesium (91 ± 0.027), potassium (1943 ± 0.97), phosphorus (429 ± 0.50), copper (0.361 ± 0.16), manganese (1.63 ± 0.72) and zinc (2.72 ± 0.35), the results of which are expressed in mg.100g^{-1} . Although important, most of the studies found in the literature only mention guava as a good source of minerals, but do not determine them. The few studies found include Fernandes, 2007 (417 mg.100 g^{-1}), Freire et al., 2012 (1405 mg.100 g^{-1}) and Xavier, 2013 (14.20 mg.100 g^{-1}), for potassium; and Oliveira et al., 2012 for calcium (14 mg.100g^{-1}) and phosphorus (30 mg.100g)$^{-1}$

4.2 Influence of adding hydrocolloids and enzymes to guava nectars

4.2.1 pH and Total Titratable Acidity (ATT)

The pH of the nectar formulation with xanthan (T2), guar (T3) and rice flour (T4) did not differ significantly from the control (T1) (4.17 ± 0.01), while the nectar with the combination of xanthan and guar (T5) and with xanthan and rice flour (T6) differed significantly, with a higher pH value on average of 4.29 ± 0.04 (Appendix A; Fig.11). The nectar formulations with guar and rice flour (T7), with the enzyme (T8), with enzyme and xanthan (T9), with enzyme and guar (T10) and with enzyme and rice flour (T11) had a lower pH value. With the exception of the formulation containing guar and rice flour (T7), all the nectars that had a lower pH than the control contained the enzyme pectinase in their formulation.

Godoy (1997), in his study of hydrocolloids, found an increase in the pH value compared to the control (guava nectar without stabilizer) when he added xanthan gum at concentrations (0.07, 0.12 and 0.17%) to guava nectar. In peach nectars, Souza (2009) also observed an increase in the pH value when xanthan and guar were added, compared to the control. However, these results do not fully corroborate those of the present study, as a significant increase in pH was only observed in the nectars added with xanthan and guar (T5) and with xanthan and rice flour (T6), and in the formulations that used xanthan and guar gums alone, no significant increase in pH value was observed. Despite the significant increase in pH, in some formulations the nectars remained within the acidic range, pH ≤ 4.5. Determining pH is important because its value is a limiting factor for the growth of pathogenic and spoiling bacteria; and therefore, to define the heat treatment to be applied; in addition to favoring the stability of ascorbic acid, since this vitamin has greater stability at acidic pH (ZAMBIAZI, 2010). According to Evangelista et al. (2006), each microorganism has a minimum and maximum pH range for growth, with most of them growing at pH levels close to neutrality (6.6-7.5) and few being able to develop at pH values below 4.0; therefore, the nectars formulated in this study have a pH at which pathogenic microorganisms are unlikely to grow.

During the storage period, there was a significant increase in the pH value of all the nectar formulations, with the exception of the nectar added with xanthan and rice flour (T7), with guar (T3) and with xanthan (T2) at 180 days of storage. These results corroborate those of Silva (2010), who assessed the stability of tropical guava juice bottled in glass and carton packaging at room temperature for 250 days; Mattietto (2007), who studied the stability of mixed cajá and umbu nectar bottled in glass bottles during 90 days of storage; Leitão (2007), who assessed the stability of blackberry nectar stored in glass and polypropylene packaging at room temperature and under refrigeration; and Carvalho et al. (2007), assessing the stability of a mixed drink containing cashew juice and coconut water with added caffeine. The increase in pH value, according to Silva (1999), is due to the degradation of ascorbic acid during storage, with a corresponding reduction in free hydrogen ions in the mixture. This corroborates the results of this study (figure 21; Appendix C), where a significant decrease in the vitamin C content can be seen over the course of storage.

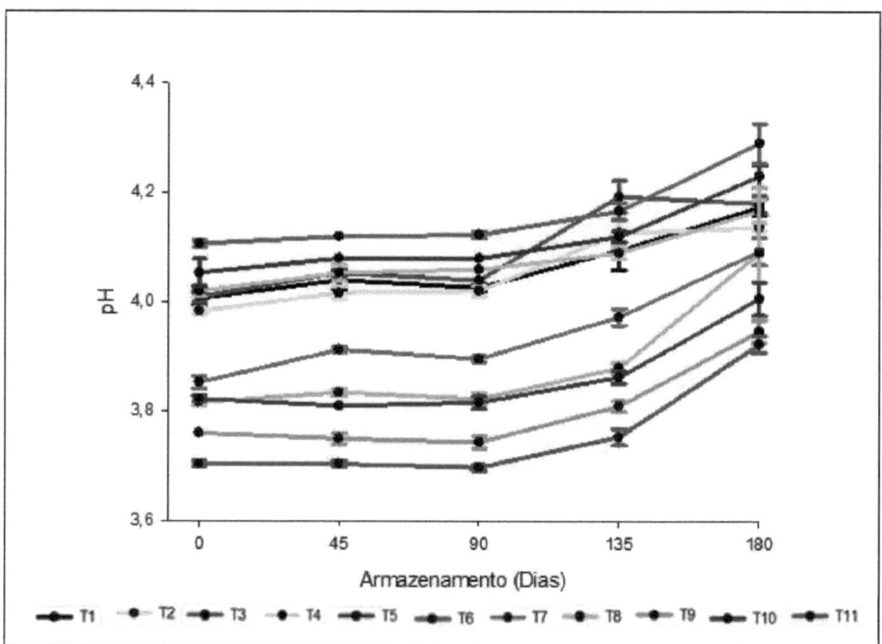

Figure 11. Influence of the addition of hydrocolloids and enzymes on the pH of guava nectars during storage. T1-control, T2-xanthan 0.1%, T3-guar 0.1%, T4-rice flour 0.1%, T5-xanthan 0.05% + guar 0.05%, T6-xanthan 0.05% + rice flour 0.05%, T7-guar 0,05% + rice flour 0.05%, T8-enzyme 1400 ppm, T9-enzyme 1400 ppm + xanthan 0.1%, T10-enzyme 1400 ppm + guar 0.1%, T11-enzyme 1400 ppm + rice flour 0.1%.

Determining acidity is another important physicochemical parameter for processing nectars, as knowing the acidity guarantees a more pleasant taste and a more vivid color for the products. Among the nectar formulations that showed higher total titratable acidity values immediately after processing, differing significantly from the control, were the nectars with guar and rice flour (T7) and all the nectars containing enzymes: with enzyme (T8), with enzyme and xanthan (T9), with enzyme and guar (T10) and with enzyme and rice flour (T11) (Appendix A; Figure 12). The same trend, with the same formulations, was observed in relation to the lowest pH value.

Essa (2002), evaluating the effects of using an enzyme preparation on plum, banana and guava juices, found an average titratable acidity of around 0.33% (malic acid), a value very close to that found in the present study in nectars with added enzyme (0.27% citric acid). However, the author did not observe any increase in acidity after the enzyme treatment, only a considerable reduction in the viscosity of the banana juice. Byaruagaba-Bazirake et al. (2012) also observed no changes in the titratable acidity of fruit pulps as a result of the enzymatic treatment used. The decrease in pH value and increase in titratable acidity in the enzyme-infused nectar formulations is justifiable, since the enzyme treatment increases the galacturonic acid content in the medium. The increase corresponds to the extraction of this acid present in the form of pectin chains in the cell walls. An increase in acidity was also observed by Demir et al. (2004), although the researchers did not observe a change in pH as occurred in the present study, they believe that some components of the carrot juice acted as a buffer solution, which would justify the absence of a change in pH in their results. Vandresen (2007), evaluating enzymatically treated and pasteurized carrot juice, also observed a decrease in the pH value and an increase in the titratable acidity of the samples.

It was observed that the total titratable acidity remained practically stable for all the formulations during the 180 days of storage, but did not follow its reduction as a result of the increase in pH value, especially at 135 days of storage. There was a slight decrease in acidity at some of the times and formulations evaluated, with the formulation with enzyme and xanthan (T9) being the most noticeable.

Pinheiro (2008), studying mixed cashew-based nectar during 30 days of storage, observed small oscillations in titratable acidity for nectar packaged in polyethylene terephthalate (PET) and glass containers. Similar trends were observed by Corrêa (2002), when evaluating guava nectar formulations stored at room temperature (25 ± 5°C) and refrigerated (5 ± 2°C) for 120 days; and by Gofur et al. (1994), cited by Beisman (2000), during the storage of different mango nectar formulations.

Despite the variations found for the total acidity of the different formulations of guava nectars in this study, all the values found were higher than the minimum established by the legislation (PIQ), which is 0.10 % (BRASIL, 2003).

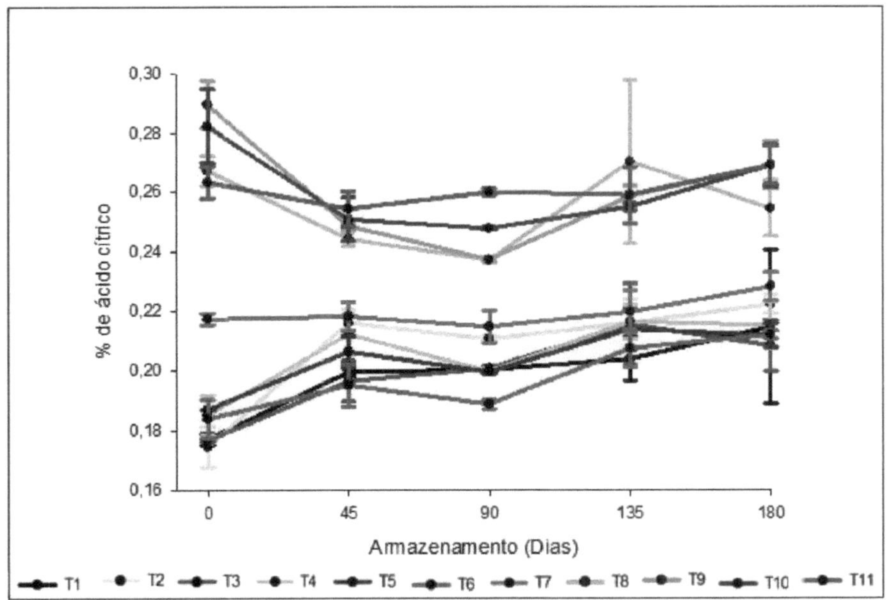

Figure 12. Influence of the addition of hydrocolloids and enzymes on the total titratable acidity of guava nectars during storage. T1 - control, T2-xanthan 0.1%, T3-guar 0.1%, T4-rice flour 0.1%, T5-xanthan 0.05% + guar 0.05%, T6-xanthan 0.05% + rice flour 0.05%, T7-guar 0,05% + rice flour 0.05%, T8-enzyme 1400 ppm, T9-enzyme 1400 ppm + xanthan 0.1%, T10-enzyme 1400 ppm + guar 0.1%, T11-enzyme 1400 ppm + rice flour 0.1%.

4.2.   2Total Soluble Solids (TSS)

The soluble solids content of the guava nectar formulations compared to the control was significantly higher (14.03 ± 0.12) (Appendix A, Figure 13), with the exception of the formulation with guar (T3), which did not differ from the control. This increase in soluble solids content was to be expected, since solids were added to all the formulations. The formulation with enzyme and rice flour (T11) had the highest soluble solids value (16.10 °Brix), with the highest values being observed in the formulations containing enzymes (T8, T9, T10, T11). This increase can be explained by the hydrolytic action of the enzymes pectinolytic enzymes in the α (1→4) bonds, which increases the amount of soluble solids in solution. According to Sreenath et al. (1994) the addition of enzymes improves the quality of the juice because it allows for the extraction of a greater quantity of soluble solids. This effect was also verified by Brasil et al. (1995) in the extraction and clarification of guava juice using 600 ppm of enzyme at 45°C for 120 minutes; and by Vandresen (2007) evaluating enzymatically treated and pasteurized carrot juice.

Throughout storage, there was no noticeable upward or downward trend in the soluble solids content of the different samples, which corroborates the study by Nisida, Menezes and Tocchini (2002), who when investigating orange juice packed in aseptic packaging and stored at different temperatures (2, 12 and 35°C), observed that the concentration of soluble solids remained constant throughout storage at the three temperatures studied. According to Pedrão et al. (1999), small reductions in soluble solids during the storage of nectars could be due to the presence of organic acids, mainly citric acid, which cause the partial degradation of sugars to form hydroxymethyl furfural (HMF) and furfural. Oliveira et al. (1984) detected this behavior when evaluating the stability of cupuaçu nectar formulations, where they observed a slight decrease in soluble solids in the last periods of storage.

According to the Ministry of Agriculture, Livestock and Supply (MAPA), the minimum soluble solids content for guava nectars is 10°Brix (BRASIL, 2003); thus, the 11 guava nectar formulations were within the minimum limit set by legislation until the end of the storage period.

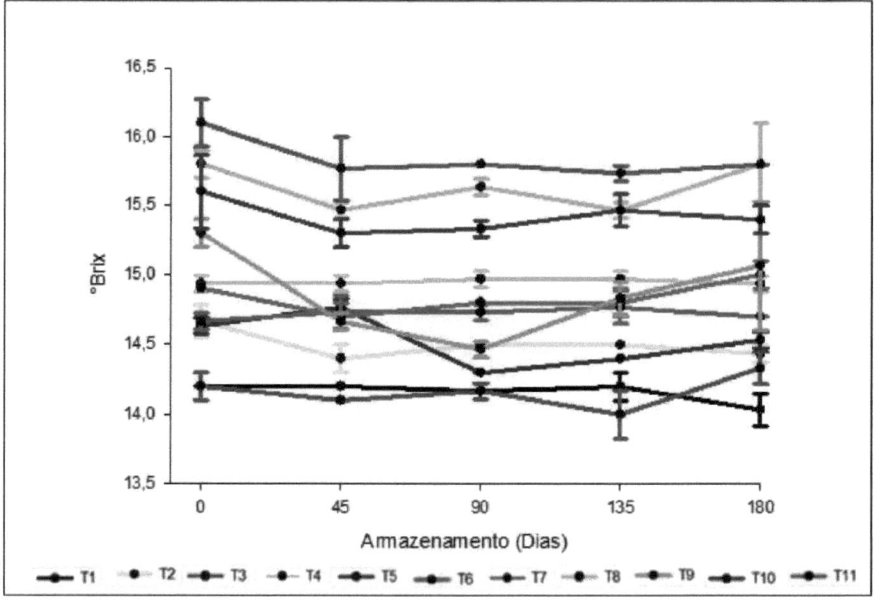

Figure 13. Influence of the addition of hydrocolloids and enzymes on the total soluble solids content of guava nectars during storage. T1 - control, T2-xanthan 0.1%, T3-guar 0.1%, T4- rice flour 0.1%, T5-xanthan 0.05% + guar 0.05%, T6-xanthan 0.05% + rice flour 0.05%, T7-guar 0,05% + rice flour 0.05%, T8-enzyme 1400 ppm, T9-enzyme 1400 ppm + xanthan 0.1%, T10-enzyme 1400 ppm + guar 0.1%, T11-enzyme 1400 ppm + rice flour 0.1%

4.2. 3 Color analysis

In addition to soluble solids content, acidity and pH, color is another important quality parameter for fruit and, consequently, its products, especially for industry. However, this quality parameter can be affected by thermal processing, which aims to reduce the initial microbial load, as well as inactivating enzymes that may cause alterations, thus increasing the product's shelf life, but which, depending on the time x temperature binomial used, can considerably affect color, due to chemical reactions and degradation of compounds such as carotenoids, vitamin C and phenolic compounds, which are directly related to the darkening of products. This is why it s essential to monitor this parameter when assessing the quality of a food or drink.

The color parameters (L*, a* and b*) of most of the nectars, at each period evaluated, remained very close to the control, and did not differ significantly (Appendix B; Figure 14); only at the beginning (0 days) and end of storage (180 days) was there a significant decrease in the L* parameter in the treatments with xanthan (T2), guar and rice flour (T7), enzyme (T8), enzyme and xanthan (T9), enzyme and guar (T10) and enzyme and rice flour (T11). Again, among the treatments responsible for the statistical difference in luminosity were those with the enzyme pectinase. However, when each nectar was evaluated individually in relation to storage time, it was observed that luminosity increased, with the greatest statistical difference at 90 days of storage, with a subsequent reduction. Saron et al. (2007), in studies on passion fruit juice, also found a significant increase in luminosity values during 120 days of storage. According to the authors, reactions such as Maillard may occur, where non-enzymatic browning occurs; or even the polymerization of phenolic compounds, which is also another factor that promotes browning. Tribst (2008), when evaluating mango nectar, also found an increase in luminosity and justified this by the heat treatment the product had undergone. Arruda (2003), in a study of the stability of mango nectars packaged in PET bottles, aluminum packaging and cartons, observed a reduction in L* during storage and attributed the darkening to oxidation reactions and the concentration of vitamin

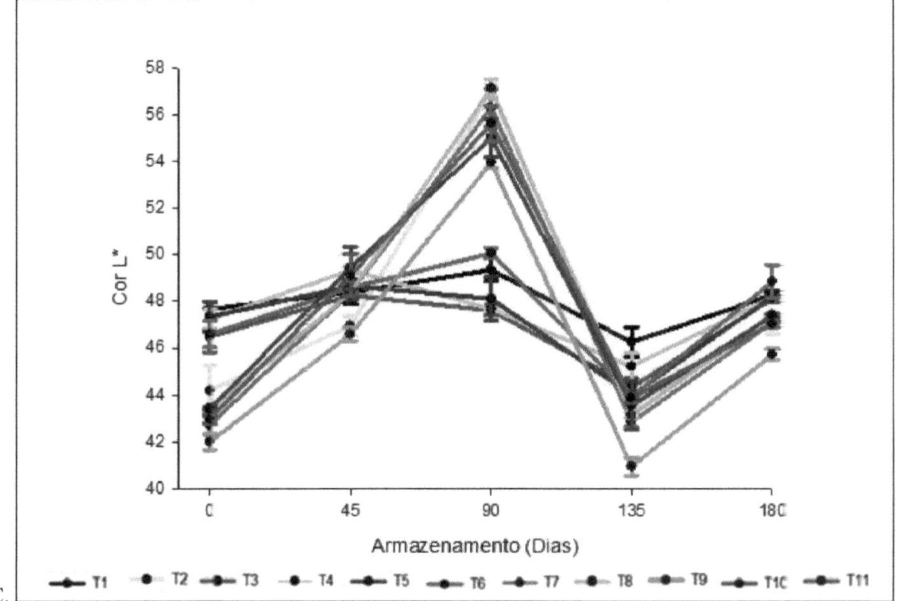

Figure 14. Influence of the addition of hydrocolloids and enzymes on the L* color parameter of guava nectars during storage. T1-control, T2-xanthan 0.1%, T3-guar 0.1%, T4-rice flour 0.1%, T5-xanthan 0.05% + guar 0.05%, T6-xanthan 0.05% + rice flour 0.05%, T7-guar 0,05% + rice flour 0.05%, T8-enzyme 1400 ppm, T9-enzyme 1400 ppm + xanthan 0.1%, T10-enzyme 1400 ppm + guar 0.1%, T11-enzyme 1400 ppm + rice flour 0.1%.

Both the data for the a* variable (Appendix B; Figure 15) and the b* variable (Appendix B; Figure 16) showed a trend towards increasing values for all the guava nectars during storage. This trend can be explained by the content of carotenoids present in the nectars, since the higher the a* and b* values, the closer they are to red and yellow hues, respectively. As observed, the different nectar formulations induced color values similar to the control.

The increase in the a* parameter was also observed by Saron et al. (2007), evaluating the stability of passion fruit juice during storage. The increase in the red hue may have been influenced by the heating of the product, which causes the loss of some components such as carotenoids, sugars and amino acids, which lead to the formation of colored products resulting from the Maillard reaction.

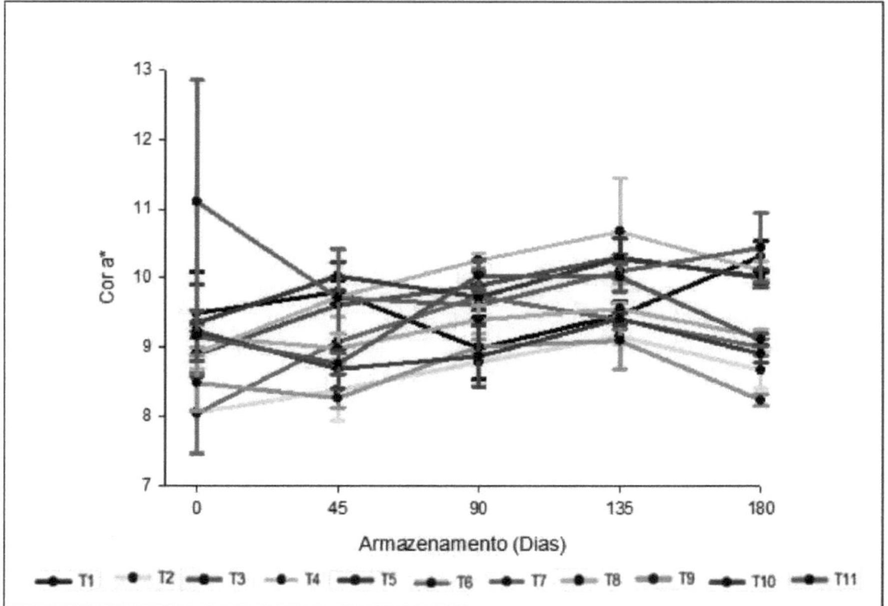

Figure 15. Influence of the addition of hydrocolloids and enzymes on the a* color parameter of guava nectars during storage. T1-control, T2-xanthan 0.1%, T3-guar 0.1%, T4-rice flour 0.1%, T5-xanthan 0.05% + guar 0.05%, T6-xanthan 0.05% + rice flour 0.05%, T7-guar 0,05% + rice flour 0.05%, T8-enzyme 1400 ppm, T9-enzyme 1400 ppm + xanthan 0.1%, T10-enzyme 1400 ppm + guar 0.1%, T11-enzyme 1400 ppm + rice flour 0.1%.

With regard to the b* variable, only some formulations showed higher values than the control at 90 and 135 days of storage, while at the other times the values did not differ statistically from the control. During the 180-day storage period, there were several oscillations in this parameter, such as a more significant decrease at 45 days, but these oscillations did not differ significantly between the different formulations and the control at the end of the evaluation period. The total variation in this analyzed parameter was between 1.24 and 5.38, thus showing a slight increase in the intensity of the yellow color during storage, which can be explained by the heat treatment applied to the nectars, which in turn causes an increase in the yellow hue of the products. Aradhita et al. (1995), cited by Beisman (2000), in a study evaluating the darkening of guava nectar, observed that the color change intensified during storage at room temperature, probably due to the rapid degradation of the ascorbic acid present in the product. It is important to know this coordinate in guava nectars, as this parameter is also directly related to the content of carotenoids in the product. As there was no great decline in the carotenoid content during storage, this was directly reflected in the values of the b* parameter.

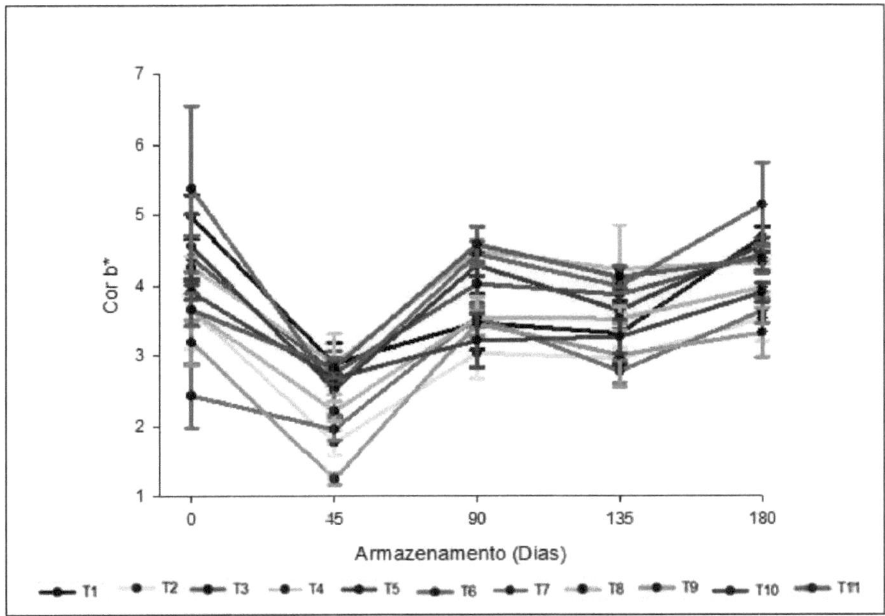

Figure 16. Influence of the addition of hydrocolloids and enzymes on the b* color parameter of guava nectars during storage. T1-control, T2-xanthan 0.1%, T3-guar 0.1%, T4-rice flour 0.1%, T5-xanthan 0.05% + guar 0.05%, T6-xanthan 0.05% + rice flour 0.05%, T7-guar 0,05% + rice flour 0.05%, T8-enzyme 1400 ppm, T9-enzyme 1400 ppm + xanthan 0.1%, T10-enzyme 1400 ppm + guar 0.1%, T11-enzyme 1400 ppm + rice flour 0.1%.

The hue angle expresses the intensity of the hue or the color itself, and by definition, starts at the +a axis (0°) which corresponds to red, which explains why the nectars, once obtained from red guava pulp, obtained values closer to 0°. Statistically, the °Hue values between the different formulations did not differ from the control, with the sole exception of the initial storage time (Appendix B; Figure 17). Immediately after processing, a significantly lower value was observed compared to the control, with the enzyme-added treatments being responsible for this difference, since their addition increases the intensity of the color, due to their ability to extract a greater quantity of pigments, which is directly related to the breakdown of pectin by the action of pectinase enzymes, thus facilitating greater extraction of compounds previously bound to the pectin structure. Jayani, Saxena and Gupta (2005) also observed this change in the color of red wine when they added pectinolytic enzymes during the maceration of the grapes.

With regard to storage time, there was practically no significant variation in the Hue angle, except at 45 days. Thus, it can be seen that the guava nectars maintained their characteristic red color during the storage period evaluated.

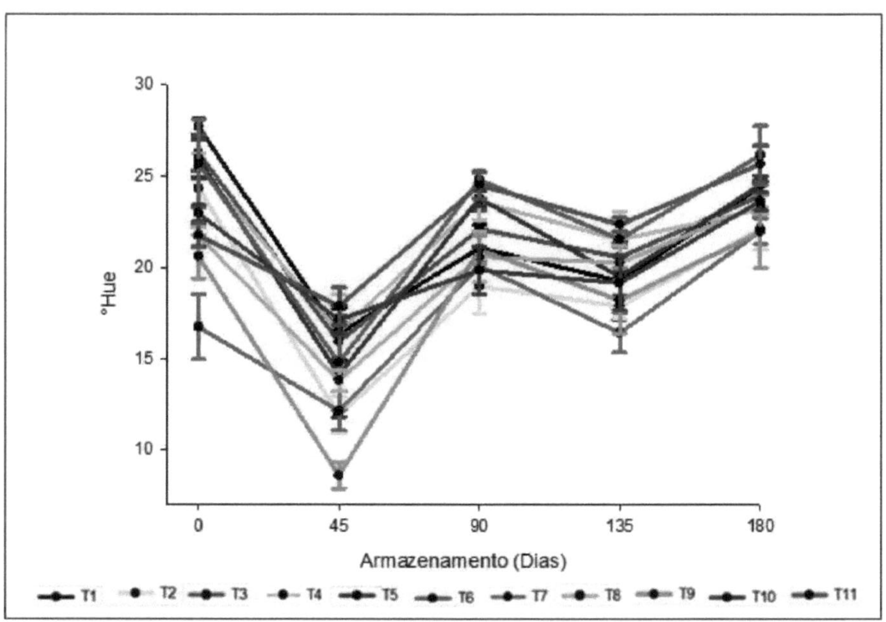

Figure 17. Influence of the addition of hydrocolloids and enzymes on the Hue values of guava nectars during storage. T1-control, T2-xanthan 0.1%, T3-guar 0.1%, T4-rice flour 0.1%, T5-xanthan 0.05% + guar 0.05%, T6-xanthan 0.05% + rice flour 0.05%, T7-guar 0,05% + rice flour 0.05%, T8-enzyme 1400 ppm, T9-enzyme 1400 ppm + xanthan 0.1%, T10-enzyme 1400 ppm + guar 0.1%, T11-enzyme 1400 ppm + rice flour 0.1%.

Chroma expresses the relationship between the a* and b* values and results in the actual color of the product analyzed, i.e. the intensity or saturation of the color. The different clarifying materials added to the nectars did not affect the color immediately after processing, with only slight variations occurring at 90 and 135 days of storage (Appendix B; Figure 18). During storage, the intensity of the color increased slightly. Jayani, Saxena and Gupta (2005), when adding pectinolytic enzymes during the maceration of grapes for red wine production, observed an improvement in color characteristics when compared to untreated wines. De Paula et al. (2004) also reported an increase in the content of carotenoids in enzymatically treated passion fruit juice, whose compounds are responsible for the variation in red, orange and yellow fruit colors. However, in the present study, there was no difference between the enzymatically treated formulations and the others.

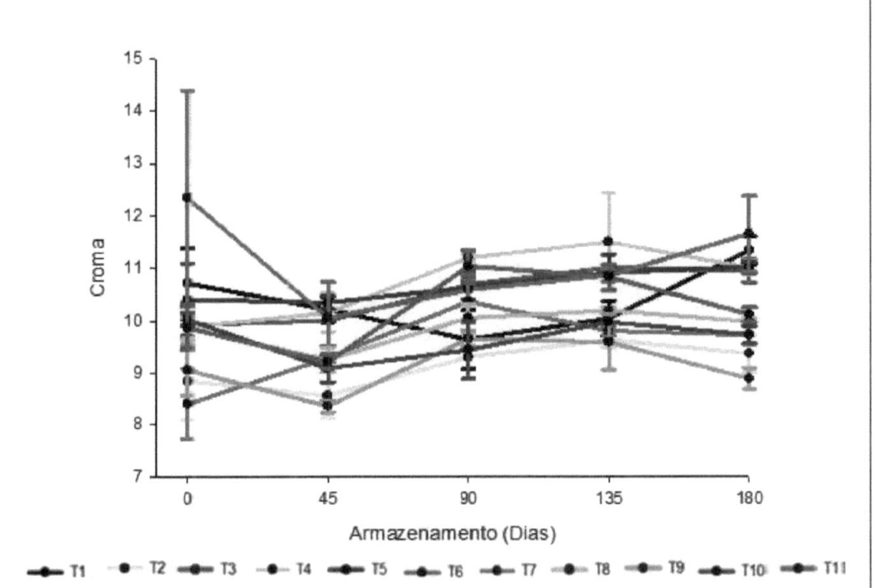

Figure 18. Influence of the addition of hydrocolloids and enzymes on the Chroma values of guava nectars during storage. T1-control, T2-xanthan 0.1%, T3-guar 0.1%, T4-rice flour 0.1%, T5-xanthan 0.05% + guar 0.05%, T6-xanthan 0.05% + rice flour 0.05%, T7-guar 0,05% + rice flour 0.05%, T8-enzyme 1400 ppm, T9-enzyme 1400 ppm + xanthan 0.1%, T10-enzyme 1400 ppm + guar 0.1%, T11-enzyme 1400 ppm + rice flour 0.1%.

4.2.4 Bioactive compounds and antioxidant capacity

4.2.4.1 Carotenoids

The carotenoid content fluctuated during the storage period (Appendix C Figure 19); however, in the xanthan sample (T2) and in the formulations containing enzymes (T8, T9, T10, T11) there was a drop (4.20mg lycopene.100g^{-1}) in the carotenoid content at the end of the storage period, while in the other samples there was even a small increase.

In studies carried out by Lin and Chen (2005), evaluating the stability of carotenoids in tomato juice heated to 121°C for 40 seconds and stored in the absence and presence of light (10 W) at 4, 25 and 35°C for 12 weeks, they observed losses in the order of 80.1; 83.5 and 82.1% for the samples stored at temperatures of 4, 25 and 35°C, respectively, in the absence of light; and 87.4; 84.9 and 88.3% at 4, 25 and 35°C, respectively, after 90 days of storage, in the samples under the effect of light. In the present study, the loss of carotenoids in the samples containing enzymes was around 19 % after 180 days of storage, i.e. a much lower drop compared to the study mentioned above. Fernandes et al. (2007), analyzing guava juice filled hot and stored for 30 days at 28 °C, observed a decrease from 1.51 mg lycopene.100g^{-1} to 1.22 mg lycopene.100g^{-1} . Silva et al. (2010) also analyzed guava juice and found a carotenoid content of approximately 1.0 mg lycopene.100g^{-1} , which did not differ when stored for 250 days. The content found in these studies is lower than that found for the guava nectars in this study.

The carotenoid content in nectar can change depending on the guava cultivar used, the extraction method, as well as oxidation, which can result from exposure to light and oxygen, the type of food matrix, the presence of enzymes, the availability of water and the presence of antioxidants and/or prooxidants (RODRIGUEZ-AMAYA, 1999). In general, the carotenoid content of the samples containing enzymes was higher than that of the other samples, both immediately after processing and at the end of the storage period. De Paula et al. (2004) also reported an increase in the carotenoid content of enzymatically treated passion fruit juice.

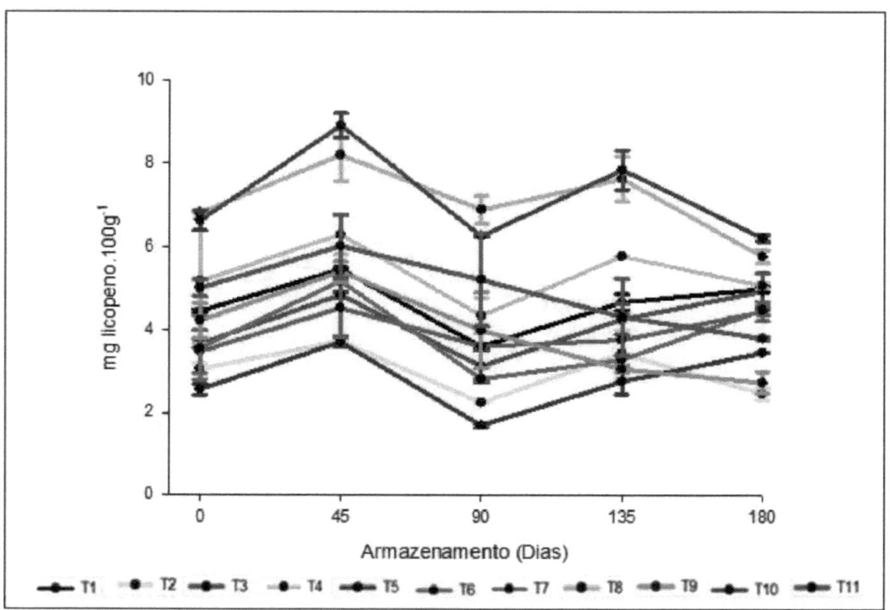

Figure 19. Influence of the addition of hydrocolloids and enzymes on the carotenoid content of guava nectars during storage. T1-control, T2-xanthan 0.1%, T3-guar 0.1%, T4-rice flour 0.1%, T5-xanthan 0.05% + guar 0.05%, T6-xanthan 0.05% + rice flour 0.05%, T7-guar 0,05% + rice flour 0.05%, T8-enzyme 1400 ppm, T9-enzyme 1400 ppm + xanthan 0.1%, T10-enzyme 1400 ppm + guar 0.1%, T11-enzyme 1400 ppm + rice flour 0.1%.

4.2.4. 2Phenolic compounds

Phenolic compounds belong to a group of substances that have variable stability, due to the different structures found in this group, and are directly affected by temperature, light, contact with oxygen, basic pH and others. All the formulations showed very similar contents of phenolic compounds immediately after processing the nectar (Appendix C; Figure 20). During the storage period, there was a significant gradual decrease in the content of phenolic compounds for all formulations, except at 90 days of storage. It can be seen that, unlike the carotenoid content, the formulations containing enzymes hardly differed from the other samples (24.89 mg gallic acid.100g^{-1}). Paludo (2011), when extracting juice with and without the enzyme pectinase, also found no significant difference in the polyphenol content of the samples. Valdés et al. (2012), in studies evaluating guava juices sold in glass packaging, found a phenol content of 26.3 mg gallic acid equivalent.100g^{-1} ; therefore, lower than in the present study. Other guava products also showed a reduction in the content of phenolic compounds during storage. Singh and Pal (2008) analyzed guava in a controlled atmosphere after storage for 30 days at 8°C, reporting decreases in the content of phenolic compounds from 224.26 mg to 190.56 mg gallic acid equivalent.100g^{-1} . Silva et al. (2010), after storing guava juice for 50 to 250 days at room temperature, found a reduction from 128.33 mg to 94.98 mg gallic acid equivalent .100g^{-1} in hot-filled juice and 96.55 mg to 74.38 mg gallic acid equivalent .100g^{-1} in aseptic-filled juice. As pasteurization was used to make the guava nectars in this study, this may be one of the causes associated wi·h the rate of degradation of phenolic compounds during storage, which was approximately 55%, but the studies cited above report even greater losses, in the order of 74 to 85%.

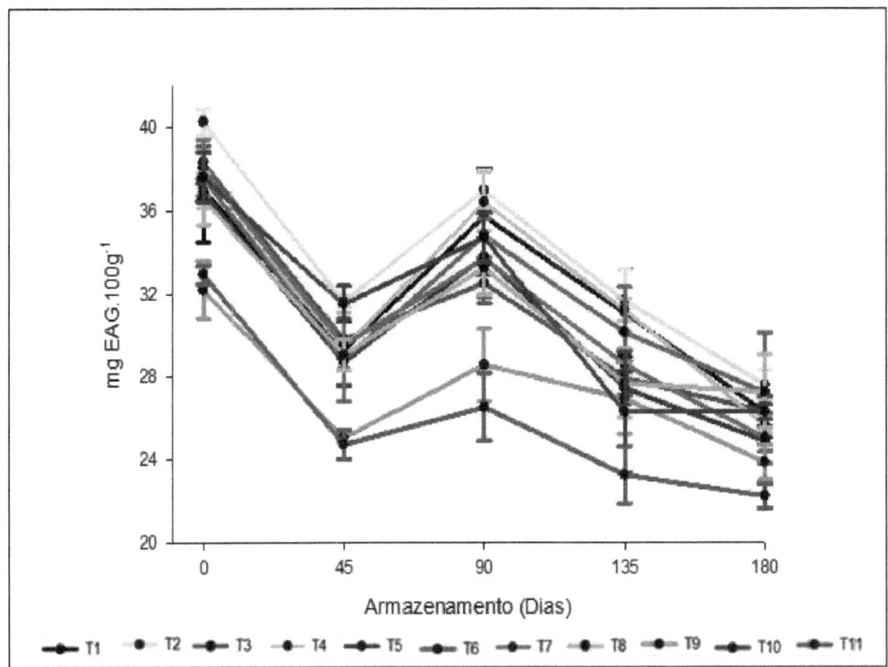

Figure 20. Influence of the addition of hydrocolloids and enzymes on the total phenolic compound content ◦f guava nectars during storage. T1-control, T2-xanthan 0.1%, T3-guar 0.1%, T4-rice flour 0.1%, T5-xanthan 0.05% + guar 0.05%, T6-xanthan 0.05% + rice flour 0.05%, T7-guar 0,05% + rice flour 0.05%, T8-enzyme 1400 ppm, T9-enzyme 1400 ppm + xanthan 0.1%, T10-enzyme 1400 ppm + guar 0.1%, T11-enzyme 1400 ppm + rice flour 0.1%.

4.2.4.3 L-ascorbic acid (vitamin C)

The content of ascorbic acid (vitamin C) in the different guava nectars was similar to that of the control nectar during all the periods evaluated, with the sole exception of the formulation containing enzyme and rice flour (T11), which had a lower content than the control throughout storage (Appendix C; Fig. 21). The vitamin C content decreased in all the guava nectars during the storage period, and at 180 days it was not detected in the formulations containing enzyme and xanthan (T9) and the one with enzyme and rice flour (T11). There was greater protection of vitamin C in the nectars with the addition of gums, except for the treatment with enzyme and guar (T10) and an acceleration in the degradation of the nectars added with the enzyme. The greatest decrease for all the formulations was observed after 135 days of storage. Leitão (2007) found that the degradation of this vitamin in blackberry nectar was 82.32% in the product stored at refrigeration temperature (4 ± 2°C) and 100% when stored at room temperature, during 90 days of storage, contributing to the assertion that juices stored at low temperatures are less prone to vitamin C degradation. Among the nectar formulations that maintained the vitamin C content throughout the storage period, in accordance with the legislation, 14 mg.100g^{-1} (BRASIL, 2003), was the treatment with xanthan (T2) (15.36 mg.100g^{-1}). Although guava is a rich source of this vitamin, the fact that it is very unstable means that it is easily degraded during processing and/or storage. Quinteros

(1995), studying the stability of acerola-carrot nectar, found faster losses in the first 90 days of storage, after which the rate of degradation decreased, unlike the present study, in which the nectars remained stable at the start of storage and degraded significantly at the end of 135 days. The tendency for vitamin C to decrease was also shown by Brito et al. (2004) during a study of the stability of passion fruit nectar made with dried coconut water, where they observed a loss of 77.87% at the end of 90 days. The same trend was observed by Sousa (2006), who reported a loss of 38% of vitamin C in his study of the stability of nectar added with extracts of Ginkgo biloba, Panax ginseng over 180 days of storage. Oliva et al. (1996), studying the stability of this vitamin in acerola nectar, reported losses of 28 to 30% when stored at room temperature after 150 days. The reduction in the vitamin C content of the nectar during storage can be attributed to oxidation reactions due to the oxygen present inside the packaging, as well as the oxygen dissolved in the drink, since the nectar did not go through the deaeration process. The storage temperature and the incidence of light on the transparent glass packaging may also have contributed to the reduction in vitamin C content (CARVALHO, 2005). According to Fellows (1997) pasteurization, as well as being a relatively mild heat treatment, also causes changes in the nutritional value of food, especially in relation to vitamin C in fruit juices. There is a vast literature that comments on the chemical oxidation of vitamin C and/or thermal degradation as a consequence of blanching, cooking, pasteurization, sterilization, dehydration and freezing (BURDURLU et al., 2006). In addition to these processing conditions, the type of packaging, the presence of O_2, the storage time/temperature ratio and the incidence of light (CORREA-NETO and FARIA, 1999) also contribute to the degree of vitamin C degradation.

Despite the significant loss of ascorbic acid in guava nectars up to 180 days of storage, in some formulations up to 90 days of storage they showed values above the Recommended Daily Intake (RDI) for adults, which is 45 mg (BRASIL, 2005).

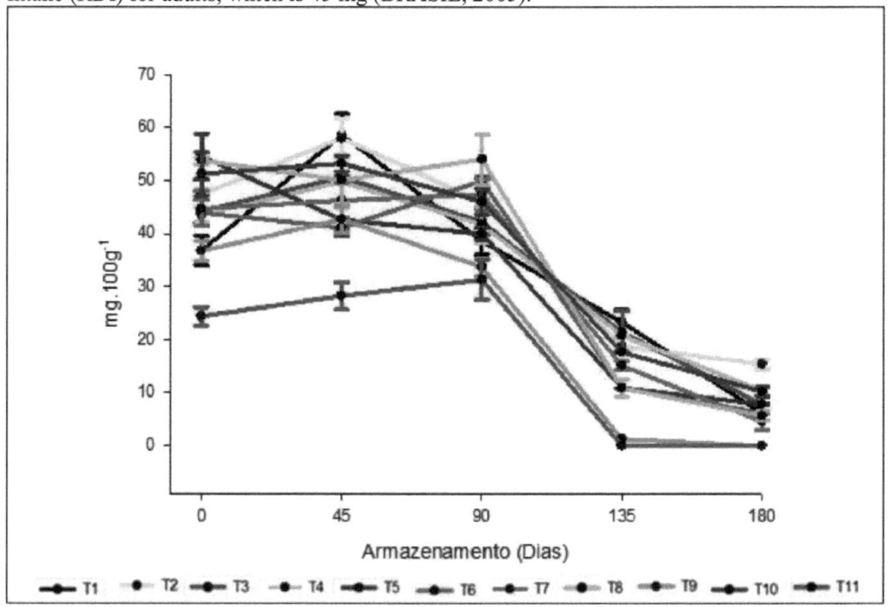

Figure 21. Influence of the addition of hydrocolloids and enzymes on the L-ascorbic acid (vitamin C) content of guava nectars during storage. T1-control, T2-xanthan 0.1%, T3-guar 0.1%, T4-rice flour 0.1%, T5-xanthan 0.05% + guar 0.05%, T6-xanthan 0.05% + rice flour 0.05%, T7-guar 0,05% + rice flour 0.05%, T8-enzyme 1400 ppm, T9-enzyme 1400 ppm + xanthan 0.1%, T10-enzyme 1400 ppm + guar 0.1%, T11-enzyme 1400 ppm + rice flour 0.1%.

4.2.4.4 Antioxidant Capacity

Analysis of the guava nectars in terms of antioxidant capacity showed that there was a greater variation in the antioxidant activity of the samples compared to the control from 90 to 135 days of storage (Appendix C; Figure 22). During storage, the samples containing the different formulations showed variable activity, where the samples containing enzymes (T8, T9, T11), except for the sample with enzyme and guar (T10), showed a tendency to reduce their antioxidant activity, which is directly related to the loss of carotenoids, phenolic compounds and vitamin C during storage, which were relatively higher in these treatments compared to the other treatments, especially in the treatments with enzyme and xanthan (T9) and enzyme and rice flour (T11), which even failed to detect vitamin C at the end of storage, a compound also responsible for antioxidant activity. The sample using only xanthan (T2) showed an increase in antioxidant capacity during the storage period, demonstrating its stabilizing potential, not only in terms of the product's appearance but also in relation to potentially bioactive compounds. Leitão (2007), evaluating the stability of blackberry nectar, found a tendency for the antioxidant capacity to increase by approximately 9% in both conditions under which the product was studied (room temperature and refrigeration). In an evaluation of guava juices sold in glass packaging, Valdés et al. (2012) found a DPPH inhibition percentage of 30%, which is lower than in the present study, in which the lowest percentage found was approximately 33% inhibition at the end of storage for the treatment containing the enzyme and rice flour (T11).

Overall, all the percentages of antioxidant activity found were considered low, given that guava is rich in phenolic compounds, carotenoids and vitamin C, which are the substances responsible for antioxidant activity. The literature shows that the acetonic and methanolic extracts of guava, at different concentrations, obtained inhibition percentages ranging from 93 to 94 and 67 to 68% respectively (ALOTHMAN, 2009). However, in this study, we found a higher percentage of inhibition in the nectar (~ 40%) compared to the pulp (~ 30%).

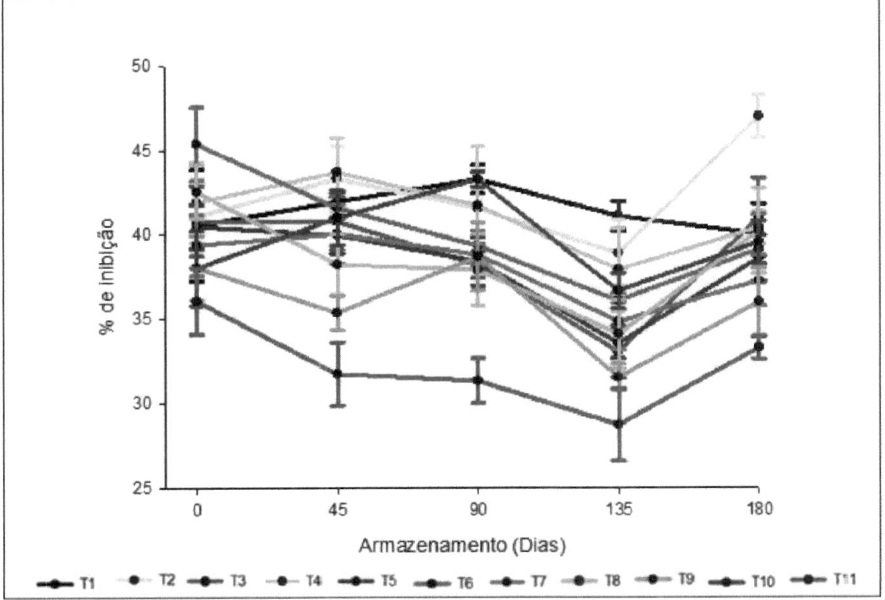

Figure 22. Influence of the addition of hydrocolloids and enzymes on the antioxidant capacity of guava nectars during storage. T1-control, T2-xanthan 0.1%, T3-guar 0.1%, T4-rice flour 0.1%, T5-xanthan 0.05% + guar 0.05%, T6-xanthan 0.05% + rice flour 0.05%, T7-guar 0,05% + rice flour 0.05%, T8-enzyme 1400 ppm, T9-enzyme 1400 ppm + xanthan 0.1%, T10-enzyme 1400 ppm + guar 0.1%, T11-enzyme 1400 ppm + rice flour 0.1%.

4.2.4. 5 Nectar stabilization test

Figure 23 shows the phase separation that occurred in the guava nectars. Sedimentation occurred more quickly at the start of the process, specifically in the first two weeks of storage, followed by a slow increase in the height of the precipitate until a stable height was reached. All the samples containing isolated or associated xanthan (T2, T5, T6), except for the sample with xanthan and enzyme (T9), did not significantly influence the stabilization of the nectar when compared to the control sample. All the other formulations showed greater stabilization than the control, especially the formulations containing rice flour (T4) and the presence of enzymes (T8, T9, T10, T11), reaching approximately 88, 80, 91 and 75% stabilization at the end of storage, The samples with added enzymes showed significantly better stabilization of the nectars compared to the samples containing the other gums, especially the samples containing only the enzyme (T8), with 100% stabilization up to the fifteenth day of storage, and the sample containing enzyme and guar (T10), which showed 100% stabilization up to the sixty-sixth day of storage.

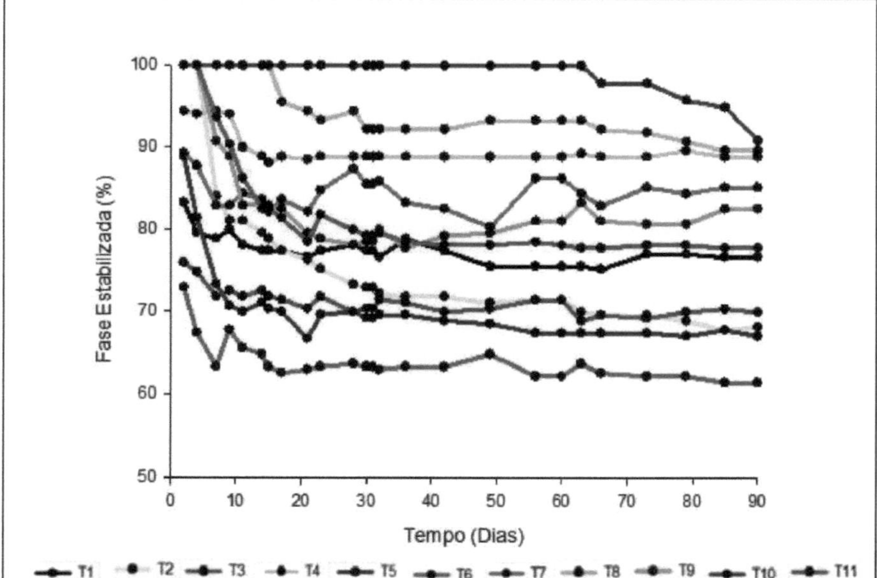

Figure 23. Influence of the addition of hydrocolloids and enzymes on the sedimentation of guava nectars during storage. T1 - control, T2-xanthan 0.1%, T3-guar 0.1%, T4-rice flour 0.1%, T5-xanthan 0.05% + guar 0.05%, T6-xanthan 0.05% + rice flour 0.05%, T7-guar 0,05% + rice flour 0.05%, T8-enzyme 1400 ppm, T9-enzyme 1400 ppm + xanthan 0.1%, T10-enzyme 1400 ppm + guar 0.1%, T11-enzyme 1400 ppm + rice flour 0.1%.

The sample containing xanthan with rice flour (T6) had the worst performance in terms of nectar stabilization. It should be noted that the sample containing only rice flour (T4) showed a high percentage of stabilization, thus inferring that the xanthan used at the concentrations in this study did not play an important role in the stabilization of guava nectars. Godoy (1997), evaluating the stabilization of guava nectar during 180 days of storage, achieved 99% stabilization using 0.175% xanthan gum, a slightly higher amount than that used in the present study. Garruti (1989), with the addition of 0.2% xanthan gum, stabilized 100% of passion fruit juice for a period of 180 days, while

Souza (2009) also highlights the addition of 0.2% xanthan gum as one of the best treatments for stabilizing peach nectar (94.7%). However, it can be seen that in this study all the formulations that used xanthan gum performed poorly as a stabilizing agent, even when combined with enzyme, which in general showed the best results.

The fact that xanthan gum did not show good results in terms of stabilizing guava nectars in this study was not in line with what has been described in the literature, since this gum is known to be excellent for stabilizing products, and its effect is mainly associated with the characteristics of its molecule, which has a high molecular weight and many branches, which contribute to increased interactions with the compounds in the product, generating an increase in the viscosity of the dispersing medium, thus reducing the sedimentation speed of the particles. However, in relation to the presence of enzymes, Vendrúscolo (2005) observed that enzymatic treatment of carambola pulp reduces sedimentation by around 62% compared to untreated pulp, probably due to the solubilization of pectin and the release of intercellular juice.

4.2.4.6 Viscosity

The viscosity results of guava nectar formulations with added hydrocolloids and/or enzymes are presented in the form of a graph of pseudoplastic behavior (Figs. 24 and 25) and the intensity of pseudoplasticity and point viscosity (Tables 5 and 6). Pseudoplasticity is characterized by a decrease in viscosity as the rate of deformation increases (CROSS, 1965; NAVARRO, 1997; SANDERSON, 1981).

Table 5: Apparent viscosity (mPa.s) at 25°C of guava nectars with added hydrocolloids and enzymes, after 45 days of storage.

Strain rate

Treatment	$10\ s^{-1}$	$30s^{-1}$	$60\ s^{-1}$	$100s^{-}$
T1	90,73	292,6	595,5	999,6
T2	90,20	292,3	594,9	999,1
T3	89,75	292,4	595,5	999,5
T4	90,28	292,1	595,3	999,2
T5	90,93	292,7	595,4	999,3
T6	90,40	292,4	595,7	999,6
T7	90,02	292,2	595,0	999,4
T8	89,87	291,7	594,7	998,9
T9	90,57	292,3	595,8	999,4
T10	89,94	292,5	594,8	999,2
T11	90,14	291,7	594,9	999,6

Table 6. Apparent viscosity (mPa.s) at 25°C of guava nectars with added hydrocolloids and enzymes, after 135 days of storage.

Strain rate

Treatment	10 s^{-1}	30s^{-1}	60 s^{-1}	100s^{-1}
T1	90,55	292,4	595,8	999,6
T2	90,54	292,3	594,7	999,0
T3	89,99	293,2	595,7	999,2
T4	90,78	292,3	595,4	999,2
T5	90,20	292,3	595,5	999,7
T6	89,97	292,6	594,8	999,6
T7	90,56	292,1	595,6	999,5
T8	90,01	291,6	595,1	998,9
T9	89,93	293,4	595,8	999,3
T10	89,82	291,3	595,1	999,2
T11	90,51	293,3	596,3	1000

It can be seen that all the samples showed pseudoplastic behavior, with only a small variation in viscosity between the different samples. After 45 and 135 days of storage, all the samples containing enzymes (T8, T9, T10 and T11) were less viscous than the samples containing gums. As the storage period progressed, the viscosity of the samples increased. This variation in viscosity between the formulations can be attributed to the different levels of sucrose inversion that occur during the preparation of the nectar, or the partial degradation of the thickener during processing, or the formation of more stable chemical bonds at the end of storage. A direct relationship was observed between the pH and the viscosity of the nectars. The nectars with the lowest pH values had the lowest viscosity, including those with enzyme (T8), enzyme and xanthan (T9), enzyme and guar (T10) and enzyme and rice flour (T11).

The formulations with xanthan (T2), guar (T3) and xanthan and guar (T5) showed the highest viscosity in the nectars at 45 and 135 days of storage. Literature data corroborates these results, as it cites xanthan as a gum that induces high viscosities even at low concentrations and reports that the combination of xanthan and guar results in more viscous solutions with high pseudoplasticity (ROCKS, 1971; DEA; MORRISON, 1975), which was indeed the case in this study. In nectars, pseudoplastic behavior is fundamental, as it keeps the nectars in suspension, preventing deposition and making their appearance attractive for consumption.

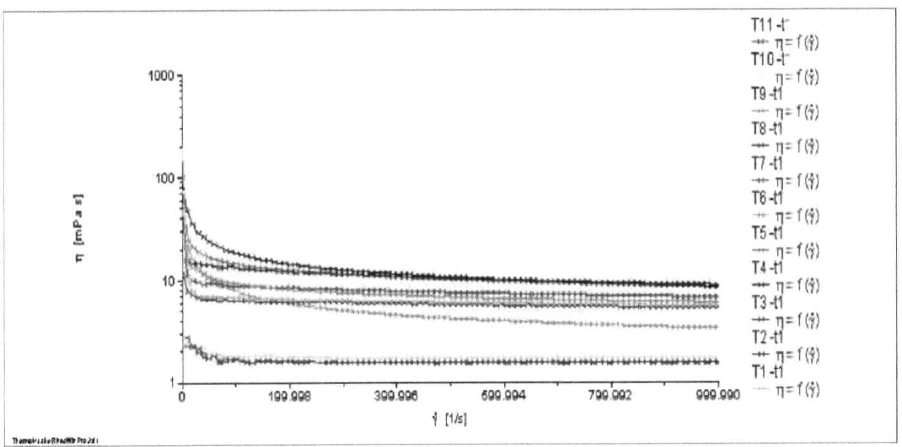

Figure 24. Rheological behavior (mPa.s) at 25°C of guava nectars with added hydrocolloids and enzymes, after 45 days of storage. T1 - control, T2-xanthan 0.1%, T3-guar 0.1%, T4-rice flour 0.1%, T5-xanthan 0.05% + guar 0.05%, T6-xanthan 0.05% + rice flour 0.05%, T7-guar 0,05% + rice flour 0.05%, T8-enzyme 1400 ppm, T9-enzyme 1400 ppm + xanthan 0.1%, T1 0-enzyme 1400 ppm + guar 0.1%, T11-enzyme 1400 ppm + rice flour 0, 1%. t1-45 days of storage.

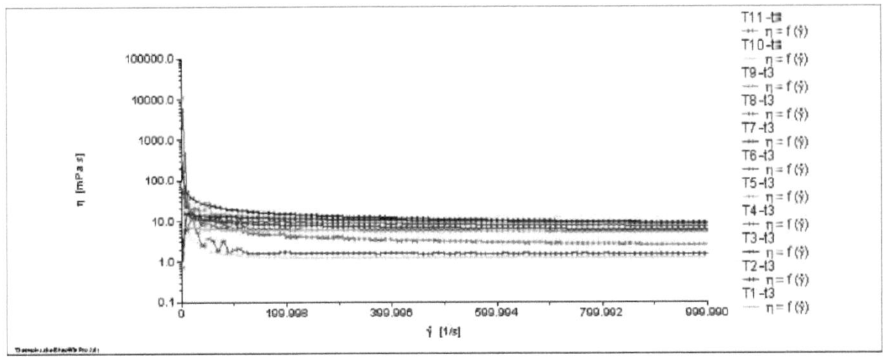

Figure 25. Rheological behavior (mPa.s) at 25°C of guava nectars with added hydrocolloids and enzymes, after 1 35 days of storage. T1 - control, T2-xanthan 0.1%, T3-guar 0.1%, T4-rice flour 0.1%, T5-xanthan 0.05% + guar 0.05%, T6-xanthan 0.05% + rice flour 0.05%, T7-guar 0,05% + rice flour 0.05%, T8-enzyme 1400 ppm, T9-enzyme 1400 ppm + xanthan 0.1%, T10-enzyme 1400 ppm + guar 0.1%, T11-enzyme 1400 ppm + rice flour 0.1%. t3-135 days of storage.

Conclusion

Guava pulp from the Paluma cultivar was found to be a rich source of carotenoids and L-ascorbic acid (vitamin C), as well as mineral salts such as calcium, magnesium, potassium and phosphorus.

The quality and physical-chemical composition of the nectars were influenced by the addition of hydrocolloids and/or enzymes.

The enzyme-added nectars allowed for greater extraction of soluble solids and carotenoids, as well as better stabilization in phase separation (75 to 91%) and less viscous samples, with the enzyme and guar treatment standing out among them.

The gum-added nectars, on the other hand, had a better effect on phenolic compounds and L-ascorbic acid (vitamin C), which directly influenced their antioxidant capacity compared to the enzymatic nectar treatments, especially the xanthan treatment.

According to the results found, it can also be concluded that it is feasible to produce these nectar formulations on an industrial scale and that the application of gums or enzymes depends on the desired objective, because while one increases the carotenoid content, the other protects the phenolic compounds and L-ascorbic acid (vitamin C).

References

ALKORTA, I.; GARBISU, C.; LLAMA, M. J.; SERRA, J. L. Industrial Application of pectic enzymes: a review. **Process Biochemistry.** v. 33, n.21, 1998.

ALOTHMAN et al. Antioxidant capacity and phenolic content of selected tropical fruits from Malaysia, extracted with different solvents. **Food Chemistry,** v. 115, p. 785-788, 2009.

AMBRÓSIO, C. L. B.; CAMPOS, F. de A. C e S.; FARO, Z. P. de. Carotenoids as an alternative to hypovitaminosis A. **Revista de Nutrição,** Campinas, v. 19, n. 2, 2006.

AMIN, A.; ESSA, H. Effect of pectinase enzyme treatment on the rheological, physical and chemical properties of plum, banana and guava juices. **Food Science and Nutrition.** v. 3, p. 13-19, 2002.

AQUINO, A. C. **Efficiency of enzymatic maceration of bacuri pulp (_Platonia insignis_ Mart.).** 2008, 129p. Dissertation (Master's in Food Science and Technology) Federal University of Ceará, 2008.

ARRUDA, A. F. **Study of the stability of mango nectar (Mandifera indica L.) bottled in PET bottles, compared to carton packs and aluminum cans.** 2003. 94p. Dissertation (Master's in Food Technology). State University of Campinas, Campinas, 2003.

ARTHEY and ARHURST. Fruit Processing: Fruit and human nutrition. New York: **Blackey Academic & Professional.** p. 20-39,1998.

ASKAR, A.; TREPTOW, H. Cloud-stable premium nectars made from tropical fruits. **Confructa,** v.36, p.130-153, 1992.

BRAZILIAN ASSOCIATION OF SOFT DRINKS AND NON-ALCOHOLIC BEVERAGES INDUSTRIES **(ABIR).** The sector. Brasília, DF, 2011. Available at: <http://abir.org.br/categoria/o-setor/>. Accessed on: November 25, 2014.

ASSOCIATION OF OFFICIAL ANALYTICAL CHEMISTS - AOAC. **Official methods of analysis.** 16.ed., Washington, 1141p., 1995.

AZZOLINI, M.; JACOMINO, A.P.; BRON, I. U. Indexes to evaluate post-harvest quality of guavas at different stages of ripeness. **Pesquisa Agropecuária Brasileira,** Brasília, v.39, n.2, p.139-145, 2004.

BARBOSA, M. M. **Obtaining carotenoids and flavonoids from cashew stalk pomace by**

enzymatic maceration. 2010, 110p. Dissertation (Master in Chemical Engineering), Federal University of Ceará, 2010.

BEISMAN, R. B. **Processing and quality evaluation of nectar and light nectar from two peach cultivars adapted to the subtropical climate.** 2000, 107p.
Dissertation (Master's Degree in Food Science and Technology), Luiz de Queiroz College of Agriculture, 2000.

BHAT, M.K. Cellulases and related enzymes in biotechnology. **Biotechnology Advances.** v. 18, p. 355-383, 2000.

BIALVES, T. S.; ARAUJO, V. F.; VIZZOTTO, M.; KROLOW, A. C. R.; FERRI, N. M. L.; NACHTIGAL, J. C. Physicochemical and functional evaluation of guava (Psidium guajava L.) cultivar paluma at different stages of ripeness. In: **Food Safety Symposium,** 4, Proceedings, Gramado, 2012.

BIBLE, B. B.; SINGHA, S. Canopy position influences CIELab coordinates of peach color. **Hortscience,** v.28, p. 992-993, 1997.

BOBBIO, P. A.; BOBBIO, F. Q. **Introduction to Food Chemistry.** 2.ed. São Paulo: Varela, 223p., 1992.

BRAND-WILLIANS, W.; CUVELIER, M.E.; BERSET, C. Use of a free radical method to evaluate antioxidant activity. **Food Science and Technology,** v.28, p. 25-30, 1995.

BRASIL, I. M.; MAIA, G. A.; FIGUEIREDO, R. W. Physical-chemical changes during extraction and clarifi cation of guava juice. **Food Chemistry,** London, v. 54, n. 4, p. 383-386, 1995.

BRAZIL. National Health Surveillance Agency. ANVISA. Ministry of Health. Resolution RDC no. 269, of September 22, 2005. Approves the technical regulation on the Recommended Daily Intake (RDI) of protein, vitamins and minerals. **Diário Oficial da União,** Poder Executivo, Brasília, September 23, 2005.

BRAZIL. Ministry of Agriculture, Livestock and Supply - Normative Instruction No. 12, of September 4, 2003. Identity and Quality Standards for Juices and Nectars. **Official Gazette of the Federative Republic of Brazil,** Brasília, DF, September 9, 2003. Available at: <http://extranet.agricultura.gov.br/sislegis> Accessed on: 18 Apr. 2013.

BRAZIL. Ministry of Agriculture, Livestock and Supply. Decree n. 6.871, of June 4, 2009. Regulates Law no. 8.918, of 14 July 1994, which provides for the standardization, classification, registration, inspection, production and supervision of beverages. **Diário Oficial da União, Brasília,** DF, 05 jun. 2009b.
Available at: <http://extranet.agricultura.gov.br/sislegisconsulta/

consultarLegislacao.do?operacao=visualizar&id =20271>. Accessed on: 13 Nov. 2013.

BRAVO, C. E. C.; CARVALHO, E. P.; SCHWAN, R. F.; GÓMEZ, R. J. H. C.; PILON, L. Determination of ideal conditions for the production of polygalacturonase by Kluyveromyces marxianus. **Ciência e Agrotecnologia.** v.24, p. 137-152, 2000.

BRITO, I.P. et al. Passion fruit nectar made with dried coconut water (Cocos nucifera, L.). In: **Congresso Brasileiro de Ciência e Tecnologia de Alimentos,** Recife, 9., 2004.

BURDURLU, H. S.;KOCA, N.; KARADENIZ, F. Degradation of vitamin C in citrus juice concentrates during storage. **Journal of Food Engineering,** England, v.74, n. 2, p. 211-216, 2006.

BYARUAGABA-BAZIRAKE, G. W.; VAN RANSBURG, P.; KYAMUHANGIRE, W. Characteristics of enzyme-treated banana juice from three cultivars of tropical and subtropical Africa. **African Journal of Food Science and Technology,** v. 3, n. 10, p. 277-290, dez. 2012.

CARDELLO, H. M. A. B. & CARDELLO, L. Vitamin C content, ascorbate oxidase activity and sensory profile of mango (Mangifera indica L.) var. Haden, during ripening. **Food Science and Technology,** v. 18, n.2, 1998.

CARDOSO, P. C.; TOMAZINI, A. P. B.; DTRIGHETA, P. C.; RIBEIRO S. M. R.; PINHEIRO-SANT'ANA, H. M. Vitamin C and carotenoids in organic and conventional fruits grown in Brazil. **Food Chemistry,** v.126, n. 2, p. 411-416, 2011.

CARVALHO, J. M.; MAIA, G. A.; FIGUEIREDO, R. W.; BRITO, E. S.; GARRUTI, D. S. dos. Mixed drink with stimulating properties based on coconut water and clarified cashew juice. **Ciência e Tecnologia de Alimentos,** Campinas, v.25, n.4, p. 813818, 2005.

CARVALHO, J. M.; MAIA, G. A.; FIGUEIREDO, R. W.; BRITO, E. S.; RODRIGUES, S. Development of a blended nonalccholic beverage composed of coconutwater and cashew apple juice containing caffeine. **Journal of Food Quality,** v.30, p. 664-681, 2007.

CAVALINI, F. C. **Ripening indices, harvest point and respiratory pattern of** 'Kumagai' and 'Paluma' guavas. 2004. 80p. Dissertation (Master's Degree) - Luiz de Queiroz College of Agriculture, University of São Paulo. Piracicaba, 2004.

CHAVES, M. da C. V.; GOUVEIA, J. P. G.; ALMEIDA, F. A. C.; LEITE, J. C. A. Physico-chemical characterization of acerola juice. **Revista de Biologia e Ciências da Terra,** v.4, n.2, 2004.

CHITARRA, M. A. F.; CHITARRA, A. B. **Post-harvest of fruits and vegetables: physiology and handling.** 2.ed. rev. and ampl. Lavras: UFLA, 785p., 2005.

CHITARRA, M.I.F.; CHITARRA, A.B. **Post-harvest of fruits and vegetables: physiology and handling.** Lavras, FAEPE, 293 p., 1990.

CORDENUNSI, B. R.; GENOVESE, M. I; NASCIMENTO, J. R. O.; HASSIMOTTO, N. M. A.; SANTOS, R. J.; LAJOLO, F. M. Effect of temperature on the chemical composition and antioxidant activity on three strawberry cultivars. **Food Chemistry,** v. 91, p. 113 - 121, 2005.

CORRÊA, M. I. C. **Processing of guava nectar (*Psidium guajava* L. var. Paluma): volatile compounds, physical and chemical characteristics and sensory quality.** 2002. 98 p. Master's thesis. Viçosa- MG: Federal University of Viçosa, 2002.

CORREA-NETO, R.S.; FARIA, J.A.F. Factors influencing the quality of orange juice. **Ciência e Tecnologia de Alimentos,** v.19, n.1, p.153-160, 1999.

COSTA, A. F. S.; PACOVA, B. E.. Botany and varieties. In: COSTA, A. F. S.; COSTA, A. N. (eds). **Technologies for guava production.** Vitória, ES: Incaper, p. 27-56, 2003.

COUTO, M. A. L.; CANNIATTI-BRAZACA, S. G. Quantification of vitamin C and antioxidant capacity of citrus varieties. **Ciência e Tecnologia de Alimentos,** v. 30, p.15-19, 2010.

COZZOLINO, S. M. F. **Bioavailability of nutrients - Vitamin C.** São Paulo: Editora Manole, 1368p., 2005.

CROSS, M.M. Rheology of non-Newtonian fluids: a new flow equation for pseudoplastic systems. **Journal Colloid Science.** n.20, p.412-437, 1965.

DE PAULA, B., MORAES, I. V. M., CASTILHO, C. C., GOMES, F.S., DA MATTA, V. M., CABRAL, L. M. C. Improvement in the clarification efficiency of passion fruit juice by combining microfiltration and enzymatic processes. **Bulletin of the Food Research and Processing Center,** Curitiba, v. 22, n. 2, 2004.

DEA, I. C. M.; MORRISON A. Chemistry and interactions of seed galactomannans. Advances in Carbohydrate, **Chemistry and Biochemistry,** v.32, p. 241-312, 1975.

DEGÁSPARI, C. H.; WASZCZYNSKY, N. Anti-oxidant properties of phenolic compounds. **Revista Visão Acadêmica,** v.5, p.33-40, 2004.

DEMIR, N.; ACAR, J.; BAHCECI, K.S. Effects of storage on quality of carrot juices produced with lactofermentation and acidification. **European Food Research and Technology,** v. 218, p. 465-468, 2004.

DICKSON, E. Hydrocolloids at interfaces and the influence on the properties of dispersed systems.

Food Hydrocolloids, v.17, p. 25-39, 2003.

DIMITRIUS, B. Sources of natural phenolic antioxidants. **Trends in Food Science and Technology,** v. 17, p. 505-512, 2006.

DORS, C. G., CASTIGLIONI, L. G.,RUIZ, W. A. Use of rice flour in the preparation of desserts. **Vetor,** Rio Grande, v.16(1/2), p. 63-67, 2006.

DZIEZAK, J.D. A focus on gums. **Food Technology,** v.45, p.115-132, 1991.

ESSA, H. A. A. Effect of pectinase enzyme treatment on the rheological, physical and chemical properties of plum, banana and guava juices. **Polish journal of food and nutrition sciences.** v. 11-52, n.3, p. 13-19, 2002.

EVANGELISTA, M. R. et al. Evaluation of the physical and chemical quality of guava and acerola juices. **Food Hygiene,** v. 20, n. 138, p. 108-114, 2006.

FELLOWS, P. **Food processing technology: principles and practice.** Abington, England: Woodhead, 505p. 1997.

FERNANDES, A. G.; MAIA, G. A.; SOUSA, P. H. M. de; COSTA, J. M. C. da.; FIGUEIREDO, R. W. de; PRADO, G. M. Comparison of vitamin C, total carotenoid and total phenolic contents of tropical guava juice at different stages of production and influence of storage. **Alimentos e Nutrição,** Araraquara, v.18, n.4, p. 431-438, 2007.

FIORUCCI, A. R. The Importance of Vitamin C in Society Through the Ages. **Química Nova na Escola,** São Paulo, v.17, p. 03-07, 2003.

FONTANA, J. D.; MENDES, S. V.; PERSIKE, D. S.; PERACETTA, L. F.; PASSOS, M. Carotenoids: Attractive colors and biological action. **Biotechnology Science and Development,** 2000.

FOOD AND AGRICULTURE ORGANIZATION OF UNITED NATIONS **(FAO/FAOSTAT).** 2012. Available at: <http //faostat.fao.org/>. Accessed on: September 17, 2014.

FREDA, S. A. **Conventional and light guava jam (*Psidium guajava* L.): stability of bioactive compounds, sensory and microbiological quality.** 2014, 98p. Dissertation (Master's Degree in Food Science and Technology), Federal University of Pelotas, 2014.

FREIRE, J. M.; ABREU, C. M. P. de; CORRÊA, A. D.; SIMÃO, A. A.; SANTOS, C. M. dos. Evaluation of functional compounds and antioxidant activity in guava pulp flours. **Revista

Brasileira de Fruticultura, Jaboticabal, v. 34, n. 3, 2012.

FREITAS, G.A. **Informe Rural Etene,** Production and harvested area of guava in the northeast, Year IV, AEPA n° 24, p 18, 2010.

GALIOTOU-PANAYOTOU, M.; KAPANTAI, M.; KALANTZI, O. Growth conditions of Aspergillus sp. ATHUM-3482 for polygalacturonase prduction. **Applied Microbiology and Biotechnology,** Berlin, v.47, n.4, p. 425-429, 1997.

GARCÍA-OCHOA, F.; SANTOS, V. E.; CASAS, J. A.; GÓMEZ, E. Xanthan gum: production, recovery, and properties. **Biotechnology Advances,** v.18, p. 549-579, 2000.

GARRUTI, D. S. **Contribution to the study of the physical stabilization of whole passion fruit juice (*Passiflora edulis* F. flavicarpa Deg).** 1989, 194p. Thesis (Doctorate in Food Technology). State University of Campinas, 1989.

GHARRAS, H. E. Polyphenols: food sources, properties and applications - a review. **International Journal of Food Science and Technology,** v. 44, p. 2512-2518, 2009.

GLICKSMAN, Martin. Background and classification. In: GLICKSMAN, M. **Food hydrocolloids.** New York : Academic Press, v.1. p. 3-18, 1982.

GODOY, R. B. **Gums in the stability of Guava Nectar and Juice (*Psidium guayava* L.).** 1997, 52p. Dissertation (Master's Degree in Agroindustrial Science and Technology). Federal University of Pelotas, Pelotas, 1997.

GODOY, R.C.B.; ANTUNES, P.L.; ZONTA, E.P. Stabilization of guava nectar (Psidium guajava L.) with xanthan gum, carrageenan and waxy starch. **Revista Brasileira de Agrociência,** v.2, p. 105-110, 1998.

GOLDSTEIN, A. M.; ALTER, E. N.; SEAMAN, J. K. Guar gum: In: WHISTLER, R. **Industrial Gums.** 2.ed. New York: Associated Press, p.303-332, 1973.

GOMES, Fabio da S. Carotenoids: a possible protection against the development of cancer. **Revista de Nutrição,** Campinas, v. 20, n. 5, 2007.

GONZAGA NETO, L.; SOARES, J. M. **Guava Culture.** Brasflia: Embrapa - SPI, 1995.

GUMANDI, S. N.; KUMAR. D. S. Enhanced production of pectin lyase and pectate lyase by Debaryomyces nepalensis in submerged fermentation by statistical methods. **American Journal of Food Technology,** New York, v.1, n.1, p. 19-33, 2006.

GUMMADI, S.N.; PANDA, T. Purification and biochemical properties of microbial pectinases:/a review. **Process Biochemistry,** v.38, p. 987-996, 2003.

HAIDA, K. S.; BARON, A.; HAIDA, K. S.; FACI, D.; HAAS, J.; SILVA, F. J. Total phenolic compounds and antioxidant activity of two varieties of guava and rue. **Brazilian Journal of Health Sciences,** v.9, n.28, p.11-19, 2011.

HE, J.; GIUSTI, M. M. Anthocyanins: natural colorants with health-promoting properties. **Annual Review of Food Science Technology,** v.1, p.163-187, 2010.

HEERD, D. et al. Pectinase enzyme-complex production by Aspergillus spp. In solidstate fermentation: a comparative study. **Food and Bioproducts Processing,** Rugby, v.90, n. 2, p. 102-110, 2012.

HERCEG, Z.; HEGEDUSIC, V.; RIMAC, S. Influence of hydrocolloids addition on the rheological properties of whey model solutions. **Acta Alimentaria,** v.29, p. 89-103, 2000.

HUI, Y. H.; BARTA, J.; CANO, M. P.; GUSEK, T. W.; SIDHU, J. S.; SINHA, N. K. **Handbook of fruits and fruit processing.** Wiley-Blackwell, 697p., 2006.

IHA, S. M.; MIGLIATO, K. F.; VELLOSA, J. C. R.; SACRAMENTO, L. V. S.; PIETRO, R. C. L. R.; ISAAC, V. L. B.; BRUNETTI, I. L.; CORRÊA, M. A.SALGADO, H. R. N. Phytochemical study of guava (*Psidium guajava* L.) with antioxidant potential for the development of phytocosmetic formulation. **Revista Brasileira de Farmacognosia,** Curitiba, v. 18, 2008.

ADOLFO LUTZ INSTITUTE. Analytical Standards of the Adolfo Lutz Institute. **Chemical and physical methods for food analysis.** IV edition. 1st Digital Edition. São Paulo. 2008.

BRAZILIAN FRUIT INSTITUTE **(IBRAF).** 2011. Available at: <http://www.ibraf.org.br/>. Accessed on: Aug. 28, 2014.

JACOBO, C. M.; AHUMADA, L. M. T.; MALDONADO, S. H. G. Caracterización de selecciones de guayaba para el bajío de Guanajuato, México. **Agricultura Técnica en México,** Texcoco, v. 35, n. 3, 2009.

JACQUES, A. C.; PERTUZATTI, P. B.; BARCIA, M. T.; ZAMBIAZI, R. C.; Bioactive compounds in small fruits grown in the southern region of the state of Rio Grande do Sul. **Brazilian Journal of Food technology,** v.12, p. 123-127, 2009.

JAYANI, R.S.; SAXENA, S.; GUPTA, R. Microbial pectinolytic enzymes: a review. **Process Biochemistry,** London, v. 40, n. 9, p. 2931-2944, 2005.

JOSEPH, B., MINI PRIYA, R. Review on nutritional, medicinal and pharmacological properties of guava (Psidium guajava L). **International Journal of Pharma Bio Sciences,** v.2(1), p. 53-69, 2011.

KALT, W. Effects of production and processing factors on major fruit and vegetable antioxidants. **Journal of Food Science,** v. 70, n. 1, p. 11-19, 2005.

KASHYAP, D. R. et al. Applications of pectinases in the commercial sector: a review. **Bioresource Technology,** Oxford, v. 77, n. 3, p. 215-227, 2001.

KATZBAUER, K. Properties and applicatios of xantan gum. **Polymer Degradation and Stability,** v.59, p.81-84, 1998.

KIST, B.B. et al. **Brazilian Yearbook of Fruit Growing 2012.** Ed. Gazeta Santa Cruz, Santa Cruz do Sul, 128p., 2012.

KLAIC, P. M. A. **Development of an acid digestion method for determining salts in xanthan and rheological potentiation of xanthan from xanthomonas arboricola pv pruni by ion exchange.** 2010. 112f. Dissertation (Master's Degree in Agroindustrial Science and Technology) - Eliseu Maciel School of Agronomy, Federal University of Pelotas, Pelotas, 2010.

KOBLITZ, M. G. B. **Food biochemistry: theory and practical applications.** Rio de Janeiro: Guanabara Koogan, 2010.

KUSKOSKI, E. M, ASUERO, A. G, MORALES, M. T, FETT, R. Wild tropical fruits and frozen fruit pulps: antioxidant activity, polyphenols and anthocyanins. **Ciência Rural.** 4(36):1283-7, 2006.

LEE, S. K.; KADER, A. A. Preharvest and postharvest factors influencing vitamin C content of horticultural crops. **Postharvest Biology and Technology,** v. 20, p. 207220, 2000.

LEITÃO, A. M. **Physico-chemical, microbiological and sensory stability of blackberry nectar *(Rubus spp.),* Cv. Tupy, packed in polypropylene, during storage.** 2007, 64p. Dissertation (Master's Degree in Science and Technology

Agroindustrial), Eliseu Maciel School of Agronomy, Federal University of Pelotas. Pelotas, 2007.

LIMA, M. A. C. de; ASSIS, J. S. de; GONZAGA NETO, L. Characterization of guava fruit and selection of cultivars in the São Francisco sub-medium region. **Revista brasileira de Fruticultura,** Jaboticabal, v. 24, n.1, 2002.

LIN, C. H.; CHEN, B. H. Stability of carotenoids in tomato juice during storage. **Food Chemistry.** v. 90, p. 837-846, 2005.

LIU, R. H. Health benefits of fruit and vegetables are from additive and synergistic combinations of phytochemicals. **Journal Clinical Nutrition,** 78:S517-S520, 2003.

LUH, B.S.; EL-TINAY, A.H. Nectars, Pulpy Juices and Fruit Juice Blends. In: NAGY S.; CHEN, C.S.; SHAW, P.E. **Fruit Juice Processing Technology.** 3.ed. Florida: agscience, chap 14. p. 532-594, 1993.

MACHADO, E.S.V; LINHARES, I.W. Low-cost calorie concentrate. **Journal of Scientific Initiation.** Unicentro Newton Paiva, 2002.

MACHADO, S. S.; TAVARES, J. T. Q.; CARDOSO, R. L.; MACHADO, C. S.; SOUZA, K. E. P. Characterization of frozen tropical fruit pulps sold in the Recôncavo Baiano. **Revista Ciência Agronômica,** v.38, n.2, p. 158-163, 2007.

MANICA, I.; KIST, H.; MICHELETTO, E.L.; KRAUSE, C.A. Competition between four cultivars and two selections of guava. **Pesquisa Agropecuária Brasileira,** Brasília, v. 33, n. 8, p. 1305-1313, 1998.

MARIANO, F. A. C.; BOLIANI, A. C.; CORRÊA, L. S.; MOREIRA, E. R. Shelf life of guavas, cv. Sassaoka, minimally processed and stored in different packages. **Revista Brasileira de Fruticultura,** special volume, p.384391, 2011.

MARTINS, E. Purification and biochemical characterization of thermostable polygalacturonases produced by the fungus *Thermoascus aurantiacus* through submerged fermentation and solid-state fermentation. 2006. 108f (Doctorate in Biological Sciences), Universidade Estadual Paulista, 2006.

MARTINS, S.; MUSSATTO, S. I.; MARTÍNEZ-AVILA, G.; MONTAÑEZ-SAENZ, J ; AGUILAR, C. N.; TEIXEIRA, J. A. Bioactive phenolic compounds: Production and

extraction by solid-state fermentation. A review. **Biotechnology Advances,** v. 29, p. 365ĕ373, 2011.

MATTIETTO, R. A.; LOPES, A. S ; MENEZES, H. C. Stability of mixed cajá and umbu nectar. **Ciência e Tecnologia de Alimentos,** Campinas. v.27, n.3, p. 456-463, 2007.

MCCOOK-RUSSELL, K. P.; NAIR, M. G.; FACEY, P. C.; BOWEN-FORBES, C. S. Nutritional and nutraceutical comparison of Jamaican Psidium cattleianum (strawberry guava) and Psidium guajava (common guava) fruits. **Food Chemistry,** Barking, v. 134, n. 2, 2012.

MELO, E. de A.; MACIEL, M. I. S.; LIMA, V. L. A. G. de; ARAÚJO, C. R. de. Total phenolic content and antioxidant capacity of frozen fruit pulps. **Alimentos e Nutrição,** Araraquara, v. 19, n. 1, 2008.

MOLLOV, P.; MALTSCHEV, E. Physico-chemical characteristics of orange juice cloud. **Journal of Food Science and Food Agriculture,** v.21, n.5, p. 250-253, 1996.

MORAIS, F. L. de. "**Carotenoids: Biological and chemical characteristics**". 2006.
70 f. Monograph (Specialization in Food Quality) ě Centre of Excellence in Tourism, University of Brasília, Brasília, 2006.

MORI, E. E. M. **Watermelon juice [*Citrullus lanatus* (Tunberg) Matsumura and Nakai]: processing, formulation, physical, chemical, microbiological characterization and acceptability.** 1996. 120p. Dissertation (Doctorate in Food Technology) ě State University of Campinas, 1996.

NASCIMENTO, R. J. do. **Antioxidant potential of guava agro-industrial waste.** 2010. 110 f. Dissertation (Master's Degree in Food Science and Technology) ě Department of Domestic Sciences, Federal Rural University of Pernambuco, Recife, 2010.

NAVARRO, R. F. **Fundamentals of Polymer Rheology.** Caxias do Sul: University of Caxias do Sul, p.300, 1997.

NERY, T.B.R; BRANDÃO, L.V.; ESPERIDIÃO, M.C.A; DRUZIAN, J.I. Biosynthesis of xanthan gum from whey fermentation: yield and viscosity. **Química Nova,** v. 31, n. 8, p.1937-1941, 2008.

NETO, C. C. Cranberry and blueberry: evidence for protective effects against cancer and vascular diseases. **Molecular Nutrition & Food Research,** v.51, p. 652-664, 2007.

NISIDA, A. L. A. C.; MENEZES, H. C.; TOCCHINI, R. P. Stability of refrigerated orange juice (Citrus Sinensis) in aseptic packaging. **Brazilian Journal of Food Technology.** Brazil, v.5, p. 95-100, 2002.

OLIVA, P. B.; MENEZES, H. C.; FERREIRA, V. L. P.. Study of the Stability of Acerola Nectar. **Ciência e Tecnologia de Alimentos,** v. 16, n. 3, p. 228-232, 1996.

OLIVEIRA, D. da S.; AQUINO, P. P.; RIBEIRO, S. M. R.; PROENÇA, R. P. da C.; PINHEIRO-SANT'ANA, H. M. Vitamin C, carotenoids, total phenolics and antioxidant activity of guava, mango and papaya from Ceasa in the state of Minas Gerais. **Acta Scientiarum,** Maringá, v. 33, n. 1, 2011.

OLIVEIRA, I. P. de.; OLIVEIRA, L. C.; MOURA, S. F. T. de.; JÚNIOR, A. F. L.; ROSA, S. R. A. da. Guava tree cultivation: from planting to management. **Revista Faculdade Montes Belos,** v. 5, n. 4, 2012.

OLIVEIRA, M. L. de.; HOLANDA, L. F. F. de.; MAIA, G. A. Estudo da estabilidade do néctar de cupuaçu (Theobroma grandiflorum Schum). **Ciência Agronômica,** Fortaleza, v.15, p. 75-77, 1984.

OSORIO, C.; FORERO, D. P.; CARRIAZO, J. G. Characterization and performance assessment of guava (Psidium guajava L.) microencapsulates obtained by spraydrying. **Food Research International,** Essex, n. 44, 2011.

PALUDO, M. C.; KRÜGER, R. L. Action of the enzyme pectinase in the extraction of jaboticaba juice. **Arquivos de Ciência da Saúde UNIPAR,** Umuarama, v. 15, n. 3, p. 279286, 2011.

PEDRÃO, M. R.; BALEIA, A.; MODESTA, R. C. D.; PRUDÊNCIO-FERREIRA, S. H. Physical-chemical and sensory stability of natural and sweetened frozen Tahiti lemon juice. **Ciência e Tecnologia de Alimentos,** Campinas (SP), v. 19, n. 2, 1999.

PEREIRA, A. C. da S. **Quality, bioactive compounds and total antioxidant activity of tropical and citrus fruits produced in Ceará.** 2009. 122 f. Dissertation (Master's in Food Technology) ě Department of Food Technology, Federal University of Ceará, Fortaleza, 2009.

PEREIRA, F. M.; CARVALHO, C. A.; NACHTIGAL, J. C. Século XXI: New dual-purpose guava cultivar. **Revista Brasileira de Fruticultura.** Jaboticabal, v. 25, n.3, p. 498-500, 2003

PEREIRA, F. M.; NACHTIGAL, J. C. Genetic improvement of the guava tree. In: NATALE, W.; ROZANE, D.E.; SOUZA, H.A. de; AMORIM, A.A (Org.). **Guava culture - from planting to marketing.** Jaboticabal, v.2, p. 375-378, 2009.

PEREIRA, J. M. de A. T. K.; OLIVEIRA, K. A. de M.; SOARES, N. de F. F.; GONÇALVES, M. P. J. C.; PINTO, C. L. O.; FONTES, E. A. F. Evaluation of the physical-chemical, microbiological and microscopic quality of frozen fruit pulps sold in the city of Viçosa-MG. **Alimentos e Nutrição,** Araraquara, v. 17, n. 4, 2006.

PESTANA, V. R.; MENDONÇA, C. R. B.; ZAMBIAZI, R. C. Rice bran: Characteristics, health benefits and applications. **Bulletin of the Food Processing Research Center.** Curitiba, v.26, n.1, p. 29-40, 2008.

PINHEIRO, A. M. **Development of mixed nectars based on cashew (*Anacardium occidentale* L) and açaí (*Euterpe oleracea* mart.).** 2008. 76 f. Dissertation (Master's degree in food technology). Department of Food Technology, Federal University of Ceará, Fortaleza, 2008.

PORCU, O. M. **Factors influencing the composition of carotenoids in guava, acerola, pitanga and their processed products.** 2004. 131 f. Thesis. (Doctorate in Food Science) ě Faculty of Food Engineering, State University of Campinas, Campinas, 2004.

PRADO, A. **Phenolic composition and antioxidant activity of tropical fruits.**
2009. 107 f. Dissertation (Master of Science) ě Escola Superior de Agricultura "Luiz de Queiroz" da Universidade de São Paulo, Piracicaba, 2009.

BRAZILIAN FRUIT EXPORT PROMOTION PROGRAM. 2012. Available at: <http://www.brazilianfruit.org/ Pbr/Fruticultura/ Fruticultura.asp>. Accessed on: September 19, 2013.

QUINTEROS, E.T.T. **Processing and stability of acerola-carrot nectar.** 1995, 96p. Dissertation (Master's Degree in Food Science). University of Campinas, Campinas, 1995.

RAMÍREZ, A.; DELAHAYE, E. P. de. Functional properties of waxes high in dietary fiber obtained from piña, guayaba and guanábana. **Interciencia,** Rio de Janeiro, v. 34, n. 4, 2009.

RIBEIRO, E. P.; SERAVALLI, E. A. G. Química **de alimentos.** 1.ed. São Paulo: Edgard Blücher Ltda, 184p., 2004.

RIOS, A. de O.; ANTUNES, L. M. G.; BIANCHI, M. de L. P. Protection of carotenoids against free radicals generated in cancer treatment with Cisplatin. **Alimentos e Nutrição,** Araraquara, v. 20, n. 2, 2009.

ROCKS, J. K. Xanthan gum. **Food Technology,** v.25, n.5, p. 22-31, 1971.

RODRIGUES, R. D. P. **Obtaining banana nectar by enzymatic maceration of the pulp of the prata-anã variety.** 2013. 96p. Dissertation (Master's in Chemical Engineering), Federal University of Ceará, 2013.

RODRIGUEZ-AMAYA, D. B. **A guide to carotenoid analysis in foods.** Washington: ILSI Press, 64p., 1999.

RODRIGUEZ-AMAYA, D. B.; KIMURA, M.; AMAYA-FARFAN, J. **Brazilian sources of carotenoids: Brazilian table of carotenoid composition in foods.** Brasília: MMA/SBF. 100p., 2008.

RODRIGUEZ-AMAYA, D. B.; PORCÚ, O. M. Pink fleshed guava and guava products as rich sources of lycopene effects of industrial processing. **Journal of the Science of Food and Agriculture,** v.22, 22p., 2004.

ROJAS-BARQUERA, D.; NARVÁEZ-CUENCA, C. E. Determination of vitamin c, total phenolic compounds and antioxidant activity of guayaba fruits (Psidium guajava L.). Cultivated in colombia. **Química Nova,** São Paulo, v. 32, n. 9, 2009.

ROSA, S.E.S. et al. Panorama of the beverage sector in Brazil. **BNDES Setorial,** n.23, p.101-150, mar. 2006. Available at: <http://www.bndes.gov.br/SiteBNDES/export/sites/default/bndes_pt/Galerias/Arquivos/ knowledge/bnset/set2304.pdf>. Accessed on: May 24, 2013.

SAKAY, T.; SAKAMOTO, T.; HALLAERT, J.; VANDAMME, E.J. Pectin, pectinase and protopectinase: production, properties and applications. **Advances in Applied Microbiology,** v.39, p.231-294, 1993.

SANDERSON, G.R. Polyssacharides in foods. **Food Technology.** v.35, n.5, p. 50-57. 1981.

SARON, E. S., DANTAS, S. T., MENEZES, H. C., SOARES, B. M. C., NUNES, M. F. Sensory stability of ready-to-drink passion fruit juice packed in steel cans. **Ciência e Tecnologia de Alimentos,** v. 27, p. 772-778, 2007.

SATO, A. C. K.; SANJINEZ, E. J.. CUNHA, R. L. Evaluation of the physical, chemical and sensory properties of industrialized guavas in syrup. **Ciência e Tecnologia de Alimentos,** Campinas, v.24, n.4, 2004.

SERRANO, L. A. L.; MARINHO, C. S.; RONCHI, C. P.; LIMA, I. M.;MARTINS, M.V. V.; TARDIN, F. D.Guava "Paluma" under different cultivation systems, times and intensities of fruiting pruning. **Pesquisa Agropecuária Brasileira,** Brasília, v.42, n.6, p. 785-792, jun. 2007.

SEVERO, J. **Ripening and UVC on transcriptional expression of genes involved in metabolic pathways of cell wall, phenolic compounds and flavors in strawberry.** 2009, 97f. Dissertation (Master's Degree in Agroindustrial Science and Technology). Federal University of Pelotas, Pelotas, 2009.

SILVA, A. de. P. V. da.; MAIA, G. A.; OLIVEIRA, G. S. F. de.; FIGUEIREDO, R. W. de.; BRASIL, I. M. Quality characteristics of pulpy cajá juice (Spondias lutea L.) obtained by mechanical-enzymatic extraction. **Revista Ciência e Tecnologia de Alimentos.** Campinas, v. 17, n.3, 1997.

SILVA, A.P.V.; MAIA, G.A.O.; OLIVEIRA, G.F.S.O et al. Study of the production of clarified cajá juice (Spondias tuberosa L.). **Ciência e Tecnologia de Alimentos,** Campinas, v.19 n.1, p. 33-36, 1999.

SILVA, C. R. M.; NAVES, M. G. V. Vitamin supplementation in cancer prevention. **Revista de Nutrição,** Campinas, v. 14, n. 2, p. 135-143, 2001.

SILVA, D. S.; MAIA, G. A.; SOUSA, P. H. M.; FIGUEIREDO, R. W.; COSTA, J. M. C.; FONSECA, A. V. V. Stability of bioactive components in unsweetened tropical guava juice obtained by the hot-fill and aseptic processes. **Ciência e Tecnologia de Alimentos,** v.30, n.1 p. 237-243, 2010.

SILVA, E. C. da; MAGALHÃES, C. H. de; GONÇALVES, R. A. Obtaining and evaluating the physical-chemical parameters of guava pulp (Psidium guajava L.), cultivar 'Paluma'. In: **IFMG Science and Technology Week,** 2, Bambuí, Proceedings. 2009.

SILVA, Marília L. C.; COSTA, Renata S.; SANTANA, Andréa dos S.; KOBLITZ, Maria G. B. Phenolic compounds, carotenoids and antioxidant activity in plant products. **Semina: Ciências Agrárias,** Londrina, v. 31, n. 3, 2010.

SINGH, S. P.; PAL, R. K. Controlled atmosphere storage of guava (Psidium guajava L.) fruit. **Postharvest Biology and Tecnology,** v.47, n.3, p. 296-306, 2008.

SIQUEIRA, Elisa B. **Physicochemical and Sensory Characterization of Guava Light Dough Jams.** 2006. 92 f. Dissertation (Master's Degree in Agroindustrial Science and Technology) ě Federal University of Pelotas, Pelotas, 2006.

SOUSA, M. A. C da.; YUYAMA, L. K. O.; AGUIAR, J. P. L.; PANTOJA, l. Açaí juice (Euterpe oleracea Mart.): microbiological evaluation, heat treatment and shelf life. **Acta Amazonica,** v.36(4), p. 483-496, 2006.

SOUZA, J. L. L. **Hydrocolloids in the physicochemical and sensory characteristics of peach nectar [*Prunus persica* (L) Batsch].** 2009. 96 f. Dissertation (Master's degree in Agroindustrial Science and Technology). Federal University of Pelotas, Pelotas, 2009.

SOUZA, P. H. M.; NETO, M. H. N; MAIA, G. A.; Functional components in food. **Bulletin of the Brazilian Society of Food Science and Technology,** p. 127-135, 2003.

SREENATH, H. K; SUDARSHANAKRISHANA, K. R.; SANTHANAM, K. Improvement of juice recovery from pineapple pulp/ residue using cellulases and pectinases. **Journal of fermentation and bioengineering,** v. 78, n.6, p. 486-488, 1994.

SUPERHIPER: **Magazine of the Brazilian supermarket association.** V.26, n.303, p. 18-20, 2000.

SWAIN, T.; HILLIS, W. E. The phenolic constituents of Prunus domestica L.- The quantitative analysis of phenolic constituents. **Journal of the Science Food and Agriculture.** Oxford, v. 10, n. 1, p. 63-68, 1959.

SWORN, G. Xanthan gum. In: PHILLIPS, G. O.; WILLIAMS, P. A. **Handbook of Hydrocolloids.** New York: CRC Press, p. 103-116, 2000.

TAIZ and ZEIGER. **Plant Physiology.** 3rd ed. Porto Alegre: Artmed, p. 312-333, 2004.
TASCA, A. P. W. **Effect of industrial processing to obtain guava on antioxidant compounds and color.** 2007. 121 f. Dissertation (Master's in Food Science) ě Faculty of Pharmaceutical Sciences, Universidade Estadual Paulista "Júlio de Mesquita Filho", Araraquara, 2007.

TAVARES, J. T. Q.; SANTOS, C. M. G.; TEIXEIRA, L. J.; SANTANA, R. S.; PORTUGAL, A. M. Stability of ascorbic acid in acerola pulp submitted to different treatments. **Magistra,** v. 15, n.2, 2003.

TAVARES, S. J. **Physical, chemical and sensory changes of rice flours submitted to microwave toasting.** 2010. 219f. Dissertation (Master's in Food Science and Technology) - Schoo of Agronomy and Food Engineering, Federal University of Goiás, Goiânia, 2010.

TEDESCO, M. J., GIANELLO, C., BISSANI, C. A., BOHNEN, H. & VOLKWEISS, S. J. Análises de solo, plantas e outros materiais. 2. ed. Porto Alegre, Universidade Federal do Rio Grande do Sul, **(Boletim técnico, 5),** 174p., 1995.

THAIPONG, K.; BOONPRAKOB, U.; CROSBY, K.; CISNEROS-ZEVALLOS, L.; BYRNE, D. H. Comparison of AETS, DPPH, FRAP, and ORAC assays for estimating antioxidant activity from guava fruit extracts. **Journal of Food Composition and Analysis,** San Diego, v. 19, 2006.

ALL FRUIT, 2009. **Information taken from the internet.** Available at: www.todafruta.com.br. Accessed on January 3, 2015.

TONELI, J.T.C.L.; MURR, F.E.X ; PARK, K.J. Study of the rheology of polysaccharides used in the food industry. **Revista Brasileira de Produtos Agroindustriais,** Campina Grande. v.7, n.2, p. 181-204, 2005.

TRIBST, A. A. L. **Effect of high dynamic pressure processing combined with mild heat treatment on the inactivation of *Aspergillus niger in* mango nectar.** Campinas, 2008, 188 p. Dissertation (Master in Food Technology), Faculty of Food Engineering, Federal University of Campinas, 2008.

 UENOJO, M.; PASTORE, G. M. Pectinases: industrial applications and prospects.

Química Nova, São Paulo, v. 30, n. 2, p. 388-394, 2007.

USTOK, F. I.; TARI, C.; GOGUS, N. Solid-state production of polygalacturonase by Aspergillus sojae ATCC 20235. **Journal of Biotechnology,** Amsterdam, v. 127, n. 2, p. 322-334, 2007.

VALDÉS, S. T, VAZ TOSTES, M. G, DELLA LUCIA, C. M, HAMACEK, F. R, PINHEIRO-SANT'ANA, H. M. Ascorbic acid, carotenoids, total phenolics and antioxidant activity in industrialized juices marketed in different packaging. **Revista Instituto Adolfo Lutz.** São Paulo, 71(4):662-69, 2012.

VALENTE, A.; SANCHES-SILVA, A.; ALBUQUERQUE, T. G.; COSTA, H. S. Development of an orange juice in-house reference material and its application to guarantee the quality of vitamin C determination in fruits, juices and fruit pulps. **Food Chemistry.** Available at: <www.sciencedirect.com>. Accessed on: January 14, 2014.

VANDRESEN, S. **Physico-chemical characterization and rheological behavior of carrot and** orange juices and their mixtures. 2007,134p. Dissertation (Master's in Food Engineering), Federal University of Santa Catarina, 2007.

VENDRÚSCOLO, A. T. **Rheological behavior and physical stability of carambola pulp** (*Averrhoa carambola* **L.).** 2005, 90p. Dissertation (Master's Degree in Food Engineering). Federal University of Santa Catarina, 2005.

VUONG, T.; BENHADDOU-ANDALOUSSI, A.; BRAULT, A.; HARBILAS, D.; MARTINEAU, L. C.; VALLERAND, D.; RAMASSAMY, C.; MATAR, C.; HADDAD, P. S. Antiobesity and antidiabetic effects of biotransformed blueberry juice in KKAy mice. **International Journal of Obesity,** p.1-8, 2009.

WILLATS, W. G. T. et al. Pectin: cell biology and prospects for functional analysis. **Plant Molecular Biology,** Dordrecht, v. 47, n. 1, p. 9-27, 2001.

XAVIER, D. **Development of a food product based on whole wheat flour and functional ingredients.** 2013, 188p. Dissertation (Master's in

UBOLDI EIROA, M. N. Deteriorating microorganisms in fruit juices and measures for control. **Bulletin of the Brazilian Society of Food Science and Technology,** v. 23, n.3/4, p. 141-160, 1989.

Chemical and Biochemical Process Technology). Federal Technological University of Paraná, 2013.

YADAV, S.; YADAV, P. K.; YADAV, D.; YADAV, K. D. S. Pectin lyase: A review. **Process Biochemistry,** v. 44, p. 1-10, 2009.

ZAICOVSKI, C. **Characterization of fruits native to the Southern Region of South America in terms of the presence of bioactive compounds, antioxidant activity and anti-proliferative activity against tumor cells.** 2008. 91f. Thesis (Doctorate in Agroindustrial Science and Technology) - Faculdade de Agronomia Eliseu Maciel, Universidade Federal de Pelotas, Pelotas, 2008.

ZAMBIAZI, R. C. **Physical-Chemical Analysis of Food.** 1st ed., Pelotas, University. 202p., 2010.

Appendices

APPENDIX A. Physico-chemical parameters of guava nectars with added hydrocolloids and enzymes

	pH				
	Time (days)				
Nectars	0	45	90	135	180
T1	4,00C	4,04C	4,03C	4,10B	4,17A
T2	3.98 cC	4.01 d*BC	4.02 dB	4.13 bcA	4.14 abA
T3	4.01 cdC	4.05 cB	4.04 cdBC	4.19 a*A	4.20 abA
T4	4.02 cD	4.05 cCD	4.06 bc*BC	4.09 cB	4.17 abA
T5	4.05 b*C	4.08 b*BC	4.08 b*BC	4.12 cB	4.23 abA
T6	4.10 a*C	4.12 a*BC	4.12 a*BC	4.17 ab*B	4.29 aA
T7	3.85 e*A	3.91 e*A	3.89 e*A	3.97 d*A	3.91 b*A
T8	3.80 f*B	3.83 f*B	3.82 f*B	3.88 e*B	4.09 abA
T9	3.76 g*C	3.75 h*C	3.74 g*C	3.81 f*B	3.95 bA
T10	3.82 ef*BC	3.81 g*C	3.81 f*C	3.86 e*B	4.01 abA
T11	3.70 h*C	3.70 i*C	3.70 h*C	3.75 g*B	3.92 bA

	Soluble Solids (°Brix)				
T1	14,20A	14,20A	14,17A	14,20A	14,03A
T2	14.67 d*A	14.40 deB	14.50 f*AB	14.50 cd*AB	14.43 deB
T3	14.20 eAB	14.10 eAB	14.17 hAB	14.00 eB	14.33 eA
T4	14.93 d*A	14.93 c*A	14.97 d*A	14.97 b*A	14.93 bcd*A
T5	14.63 d*B	14.77 c*A	14.30 g*C	14.40 dC	14.53 cde*B
T6	14.67 d*A	14.73 c*A	14.73 e*A	14.77 bc*A	14.70 cde*A
T7	14.90 d*AB	14.70 cd*B	14.80 e*AB	14.80 b*AB	15.00 bc*A
T8	15.80 ab*A	15.47 ab*A	15.63 b*A	15.47 a*A	15.80 a*A
T9	15.30 c*A	14.67 cd*BC	14.47 f*C	14.83 b*ABC	15.07 bc*AB
T10	15.60 bc*A	15.30 b*A	15.33 c*A	15.47 a*A	15.43 ab*A
T11	16.10 a*B	15.77 a*A	15.80 a*A	15.73 a*A	15.80 a*A

	Total titratable acidity (% citric acid)				
T1	0,18B	0.20AB	0.20AB	0,20A	0,21A
T2	0.17 eB	0.22 b*A	0.21 d*A	0.22 bA	0.22 bcA
T3	0.18 eB	0.20 cdAB	0.20 eA	0.22 bA	0.21 cA

continued

	0	45	90	135	180
T4	0.19 eC	0.21 bcAB	0.20 eBC	0.22 bA	0.22 bcA
T5	0.19 eC	0.21 bcdAB	0.20 eB	0.21 bA	0.21 bcA
T6	0.18 eC	0.20 dBC	0.19 f*C	0.21 bAB	0.21 bcA
T7	0.22 d*A	0.22 b*A	0.21 d*A	0.22 bA	0.23 bA
T8	0.27 bc*A	0.24 a*A	0.24 c*A	0.27 a*A	0.25 a*A
T9	0.29 a*A	0.25 a*CD	0.24 c*D	0.26 a*BC	0.27 a*B
T10	0.28 ab*A	0.25 a*BC	0.25 b*C	0.25 a*BC	0.27 a*AB
T11	0.26 c*A	0.25 a*A	0.26 a*A	0.26 a*A	0.27 a*A

Averages followed by different lowercase letters in the column differ from each other using the Tukey test (p<0.05) when evaluating the different treatments within each time (days) assessed. Averages followed by an asterisk (*) in the column differ from the control by Dunnett's test (p<0.05) within each time (days) evaluated. Averages followed by different capital letters in the row differ from each other using the Tukey test (p<0.05) when evaluating each treatment throughout the storage period (0 to 180 days). T1-control, T2-xanthan 0.1%, T3-guar 0.1%, T4-rice flour 0.1%, T5-xanthan 0.05% + guar 0.05%, T6-xanthan 0.05% + rice flour 0.05%, T7-guar 0,05% + rice flour 0.05%, T8- enzyme 1400 ppm, T9- enzyme 1400 ppm + xanthan 0.1%, T10- enzyme 1400 ppm + guar 0.1%, T11 - enzyme 1400 ppm + rice flour 0.1%.

APPENDIX B - Color attributes of guava nectars with added hydrocolloids and enzymes

	L*				
	Times (days)				
Nectars	**0**	**45**	**90**	**135**	**180**
T1	47,68 B	48.38 AB	49,33 A	46,25 C	48.28 AB
T2	44.17 b*C	46.94 bc*B	56.96 a*A	43.54 bcC	47.01 d*B
T3	46.48 aB	48.25 abA	47.61 e*A	44.32 abC	48.12 abcA
T4	47.25 aC	49.33 aA	47.67 e*BC	45.20 aD	48.20 abB
T5	47.34 aA	48.64 aA	48.10 eA	43.97 bcB	48.22 abA
T6	46.58 aC	48.57 aB	50.05 dA	43.91 bcD	48.85 aB
T7	42.71 bc*D	48.37 aB	56.28 ab*A	42.84 cD	47.08 d*C
T8	43.36 bc*D	48.35 aB	57.11 a*A	43.13 bcD	47.41 bcd*C
T9	41.99 c*D	46.58 c*B	53.94 c*A	40.95 d*E	45.72 e*C
T10	43.42 bc*D	49.42 aB	54.99 bc*A	43.54 bcD	47.38 cd*C
T11	42.96 bcD	49.10 aB	55.60 abc*A	43.86 bcD	47.04 d*C

	a*				

T1	9.49 AB	9.80 AB	8,99 B	9.47 AB	10,32 A
T2	8.05 bB	8.38 d*AB	8.79 eAB	9.16 efA	8.67 bc*AB
T3	8.90 bC	9.61 abcB	9.89 ab*AB	10.31 ab*A	10.00 aAB
T4	8.92 bB	9.73 abAB	10.26 a*A	10.68 a*A	10.11 aA
T5	9.35 abB	10.03 aAB	9.73 abc*AB	10.29 abc*A	10.02 aAB
T6	11.11 a*A	9.71 abA	9.61 abcdA	10.10 abcdA	10.45 aA
T7	8.04 bB	9.08 abcdA	9.73 abc*A	9.40 defA	9.01 b*A
T8	9.16 bAB	8.99 bcd*B	9.40 bcdeAB	9.56 bcdefA	9.15 b*AB
T9	8.48 bAB	8.26 d*B	9.02 cdeA	9.10 fA	8.23 c*B
T10	9.24 abAB	8.68 cd*B	8.87 deAB	9.42 cdefA	8.91 b*AB
T11	9.17 abA	8.76 bcd*B	10.04 ab*A	10.02 abcdeA	9.11 b*A

b*

T1	4,98 A	2,88 B	3,47 B	3,32 B	4 69 A
T2	3.66 bc*A	1.76 cd*B	3.03 dA	2.96 cdA	3.53 d*A
T3	4.39 abA	2.75 aB	4.02 abcA	3.87 abA	4.44 abA
T4	4.26 abA	2.88 aB	4.49 a*A	4.23 a*A	4.33 bcA
T5	4.56 abA	2.52 abcC	4.29 ab*AB	3.65 abcB	4.57 abA
T6	5.38 aA	2.56 abB	4.44 a*A	3.99 abAB	5.14 aA
T7	2.43 c*BC	1.95 bcd*C	3.56 bcdA	2.77 dB	3.64 cd*A
T8	3.65 bc*A	2.21 abc*B	3.54 cdA	3.53 abcdA	3.97 bcd*A
T9	3.19 bc*A	1.24 d*B	3.46 cdA	3.00 cdA	3.33 d*A
T10	3.91 bA	2.68 abC	3.21 dBC	3.27 bcdB	3.90 bcd*A
T11	3.66 bc*B	2.83 aC	4.58 a*A	4.12 a*A	4.33 abcA

°Hue

T1	27,67 A	16,38 D	21.08 BC	19.27 DC	24.42 AB
T2	24.33 abc*A	11.88 bc*D	18.97 dBC	17.88 bcC	22.12 bAB
T3	26,25A	15.94 abD	22.11 abcBC	20.58 abC	23.94 abAB
T4	25.54 abA	16.48 aC	23.62 ab*AB	21.54 aB	23.13 abAB
T5	25.95 aA	14.12 abC	23.79 ab*A	19.51 abcB	24.33 abA
T6	25.68 aA	14.78 abB	24.81 a*A	21.56 aA	26.15 aA

continued

T7	16.74 d*B	12.12 bc*C	20.09 cdA	16.42 cB	21.99 b*A
T8	21.73 c*AB	13.81 abC	20.59 cdB	20.27 abB	23.45 abA
T9	20.60 c*AB	8.56 c*C	20.99 bcdAB	18.20 bcB	21.97 b*A
T10	22.96 abc*A	17.15 aC	19.86 cdB	19.15 abcBC	23.64 abA
T11	21.77 bc*C	17.88 aD	24.51 a*B	22.35 a*C	25.66 aA

Chrome

T1	10.72 AB	10.22 AB	9,64 B	10,04 B	11,33 A
T2	8.85 bA	8.56 d*A	9.30 eA	9.63 d*A	9.36 cd*A
T3	9.93 bB	10.00 abcB	10.68 ab*AB	11.01 ab*A	10.94 abA
T4	9.89 bC	10.15 abBC	11.20 a*AB	11.49 a*A	11.00 aABC
T5	10.41 abA	10.34 aA	10.63 ab*A	10.92 abA	11.01 aA
T6	12.35 aA	10.05 abcA	10.59 abc*A	10.86 abcA	11.65 aA
T7	8.40 b*C	9.29 bcd*BC	10.36 abcdA	9.80 cdAB	9.72 cd*AB
T8	9.86 bA	9.25 bcd*B	10.05 bcdeA	10.19 bcdA	9.98 c*A
T9	9.06 bAB	8.36 d*B	9.66 cdeA	9.59 dA	8.88 d*AB
T10	10.03 bA	9.09 cd*B	9.44 deAB	9.97 bcdA	9.73 c*AB
T11	9.87 bB	9.21 bcd*B	11.03 a*A	10.83 abcA	10.11 bc*AB

Averages followed by different lowercase letters in the column differ from each other using the Tukey test (p<0.05) when evaluating the different treatments within each time (days) assessed. Averages followed by an asterisk (*) in the column differ from the control by Dunnett's test (p<0.05) within each time (days) evaluated. Averages followed by different capital letters in the row differ from each other using the Tukey test (p<0.05) when evaluating each treatment throughout the storage period (0 to 180 days). T1-control, T2-xanthan 0.1%, T3-guar 0.1%, T4-rice flour 0.1%, T5-xanthan 0.05% + guar 0.05%, T6-xanthan 0.05% + rice flour 0.05%, T7-guar 0,05% + rice flour 0.05%, T8-enzyme 1400 ppm, T9-enzyme 1400 ppm + xanthan 0.1%, T10-enzyme 1400 ppm + guar 0.1%, T11-enzyme 1400 ppm + rice flour 0.1%.

APPENDIX C. Bioactive compound content and antioxidant activity of guava nectars with added hydrocolloids and enzymes

	Phenolic compounds (mg gallic acid.100g⁻¹ BU)				
	Times (days) 0 45 90 135 180				
Nectars					
T1	37.05 A29	.27 C35.	73 AB31.	07 BC26.	24 C
T2	40.25 a*A31.	65 aC36.	97 aB31	.66 aC27.	56 aD
T3	38.31 abA29	.77 aC32.	53 bB 27.90 abcCD 26.30 abD		

continued

T4	37.50 abA	29.36 aB	36.41 abA	31.21 abB	25.50 abcC
T5	36.90 bA	28.66 aC	33.79 abB	27.43 abcdC	24.34 abcD
T6	37.60 abA	28.79 BC	34.81 abAB	30.10 abcBC	27.21 abC
T7	38.05 abA	29.48 aC	33.72 abB	28.55 abcCD	25.00 abcD
T8	36.58 bA	29.03 aB	33.26 abA	27.65 abcdB	27.24 abB
T9	32.20 c*A	25.03 b*BC	28.54 c*B	26.92 bcd*BC	23.34 bcC
T10	37.61 abA	31.52 aB	34.71 abAB	26.28 cd*C	26.26 abC
T11	32.97 c*A	24.69 b*BC	26.50 c*B	23.24 d*C	22.23 c*C

Carotenoids (mg lycopene.100g^{-1} BU)

T1	4,45 D	5,45 A	3,61 E	4,67 C	4,99 B
T2	3.05 cd*B	3.73 e*A	2.25 fg*C	3.44 cdef*AB	2.47 e*C
T3	3.66 cdBC	4.84 dA	3.13 defC	4.26 cdAB	4.94 bcA
T4	5.17 bAB	6.28 b*A	4.35 bcB	5.77 b*A	5.07 bAB
T5	2.57 d*B	3.71 e*A	1.69 g*C	2.77 f*B	3.46 d*A
T6	3.48 cd*B	4.53 de*A	3.61 cdeAB	3.77 cde*AB	4.46 c*A
T7	3.56 cdC	5.17 cdA	2.83 efD	3.30 def*CD	4.49 c*B
T8	6.80 a*C	8.19 a*A	6.89 a*BC	7.63 a*AB	5.76 a*D
T9	4.22 bcB	5.40 cdA	4.00 cdBC	3.07 ef*CD	2.74 e*D
T10	6.62 a*C	8.90 a*A	6.25 a*C	7.85 a*B	6.21 a*C
T11	5.00 bABC	6.01 bcA	5.21 b*AB	4.32 cBC	3.81 d*C

Ascorbic acid (mg L ascorbic acid.100g^{-1} BU)

T1	36,79 B	58,50 A	38,63 B	23,23 C	6 22 D
T2	47.38 abc*B	57.90 aA	45.98 bc*B	18.73 ab*C	15.56 a*C
T3	44.22 bcd*B	50.64 abc*A	42.31 bcB	21.49 aC	7.84 cD
T4	43.53 cd*AB	49.72 abcd*A	41.74 cB	20.82 aC	10.41 b*D
T5	54.46 a*A	42.66 cd*B	40,00 cdB	11.01 c*C	7.87 cC
T6	43.93 bcd*B	41.13 d*B	50.00 ab*A	10.96 c*C	6.20 cdD
T7	44.68 bc*A	46.27 bcd*A	47.43 abc*A	15.13 bc*B	4.60 dC
T8	53.65 a*A	50.04 abc*A	53.93 a*A	10.95 c*B	5.68 cdB
T9	36.74 dB	42.67 cd*A	33.74 deB	1.29 d*C	0 e*C
T10	51.20 ab*AB	53.14 abA	46.08 abc*B	17.69 ab*C	10.28 b*D
T11	24.42 e*B	28.31 e*AB	31.38 e*A	0 d*C	0 e*C

continued

Antioxidant activity (% inhibition)

T1	40,56 A	41,95 A	43,26 A	41,08 A	40,01 A
T2	41.08 abcB	43.30 aAB	41.54 abB	38.90 aB	46.97 a*A
T3	40.82 abcA	40.78 abA	38.22 b*A	33.03 cde*B	41.02 bA
T4	41.92 abA	43.69 aA	41.73 abA	37.93 abA	40.41 bcA
T5	40.50 abcA	39.98 abcA	38.41 ab*A	33.54 bcd*B	38.60 bcA
T6	45.37 aA	41.58 abAB	39.34 abBC	36.05 abcd*C	39.14 bcBC
T7	39.40 bcA	40.05 abA	38.83 ab*AB	34.82 abcd*B	37.25 bcdAB
T8	42.55 abA	38.24 bc*AB	37.90 b*AB	34.08 bcd*B	40.21 bcA
T9	38.06 bcA	35.35 cd*AB	38.70 ab*A	31.52 de*B	35.99 cdA
T10	38.03 bcBC	41.00 abAB	43.26 aA	36.66 abc*C	39.56 bcABC
T11	36.07 cA	31.73 d*ABC	31.31 c*BC	28.69 e*C	33.27 d*AB

Averages followed by different lowercase letters in the column differ from each other using the Tukey test (p<0.05) when evaluating the different treatments within each time period (days) evaluated. Averages followed by an asterisk (*) in the column differ from the control by Dunnett's test (p<0.05) within each time (days) evaluated. Averages followed by different capital letters in the row differ from each other using the Tukey test (p<0.05) when evaluating each treatment throughout the storage period (0 to 180 days). BU- wet basis. T1-control, T2-xanthan 0.1%, T3-guar 0.1%, T4-rice flour 0.1%, T5-xanthan 0.05% + guar 0.05%, T6-xanthan 0.05% + rice flour 0.05%, T7-guar 0,05% + rice flour 0.05%, T8-enzyme 1400 ppm, T9-enzyme 1400 ppm + xanthan 0.1%, T10-enzyme 1400 ppm + guar 0.1%, T11-enzyme 1400 ppm + rice flour 0.1%.

APPENDIX D. Percentage of stabilized phase obtained in the spontaneous sedimentation tests of guava nectar, base formulation (control) carried out over 90 days.

T1 (Control)

Days	Rep. 1	Rep. 2	Rep. 3	Average	D.P
2	81,25	75,00	87,50	81,25	6,25
4	77,50	73,75	80,00	77,08	3,15
7	81,25	70,00	77,50	76,25	5,73
9	81,25	75,00	76,25	77,50	3,31
11	78,75	72,50	75,00	75,42	3,15
14	76,25	72,50	75,00	74,58	1,91
15	73,75	75,00	75,00	74,58	0,72
17	76,25	73,75	73,75	74,58	1,44
21	76,25	71,25	76,25	74,58	2,89
23	73,75	73,75	76,25	74,58	1,44
28	75,00	72,50	78,75	75,42	3,15
30	75,00	73,75	75,00	74,58	0,72
31	75,00	73,75	75,00	74,58	0,72
32	75,00	71,25	75,00	73,75	2,17
36	76,25	73,75	78,75	76,25	2,50
42	75,00	72,50	76,25	74,58	1,91
49	72,50	70,00	75,00	72,50	2,50
56	72,50	70,00	75,00	72,50	2,50
60	72,50	70,00	75,00	72,50	2,50
63	71,25	70,00	76,25	72,50	3,31
66	71,25	68,75	76,25	72,08	3,82
73	72,50	71,25	78,75	74,17	4,02
79	72,50	71,25	78,75	74,17	4,02
85	72,50	70,00	78,75	73,75	4,51
90	72,50	70,00	78,75	73,75	4,51

Rep - repetition
SD - Standard Deviation

APPENDIX E. Percentage of stabilized phase obtained in the spontaneous sedimentation tests of guava nectar with 0.1% xanthan gum, carried out over 90 days.

T2 (Xanthan)

Days	Rep. 1	Rep. 2	Rep. 3	Average	D.P
2	100,00	100,00	100,00	100,00	0,00
4	100,00	100,00	100,00	100,00	0,00
7	90,00	76,25	80,00	82,08	7,11
9	87,50	76,25	72,50	78,75	7,81
11	87,50	78,75	70,00	78,75	8,75
14	86,25	75,00	70,00	77,08	8,32
15	85,00	75,00	68,75	76,25	8,20
17	82,50	73,75	67,50	74,58	7,53
21	80,00	73,75	66,25	73,33	6,88
23	80,00	72,50	63,75	72,08	8,13
28	77,50	68,75	63,75	70,00	6,96
30	77,50	68,75	62,50	69,58	7,53
31	77,50	68,75	62,50	69,58	7,53
32	76,25	67,50	62,50	68,75	6,96
36	75,00	67,50	62,50	68,33	6,29
42	75,00	67,50	62,50	68,33	6,29
49	73,75	67,50	61,25	67,50	6,25
56	73,75	67,50	62,50	67,92	5,64
60	73,75	67,50	62,50	67,92	5,64
63	72,50	65,00	61,25	66,25	5,73
66	71,25	65,00	61,25	65,83	5,05
73	71,25	65,00	61,25	65,83	5,05
79	70,00	63,75	61,25	65,00	4,51
85	68,75	62,50	60,00	63,75	4,51
90	68,75	62,50	61,25	64,17	4,02

Rep - repetition
SD - Standard Deviation

APPENDIX F. Percentage of stabilized phase obtained in the spontaneous sedimentation tests of guava nectar with 0.1% guar gum, carried out over 90 days.

T3 (Guar)

Days	Rep. 1	Rep. 2	Rep. 3	Average	D.P
2	68,75	75,00	75,00	72,92	3,61
4	68,75	73,75	72,50	71,67	2,60
7	67,50	68,75	68,75	68,33	0,72
9	71,25	68,75	67,50	69,17	1,91
11	70,00	68,75	66,25	68,33	1,91
14	72,50	68,75	66,25	69,17	3,15
15	71,25	67,50	66,25	68,33	2,60
17	71,25	67,50	65,00	67,92	3,15
21	70,00	65,00	65,00	66,67	2,89
23	71,25	68,75	65,00	68,33	3,15
28	68,75	65,00	65,00	66,25	2,17
30	72,50	62,50	65,00	66,67	5,20
31	72,50	62,50	65,00	66,67	5,20
32	72,50	65,00	66,25	67,92	4,02
36	72,50	65,00	65,00	67,50	4,33
42	68,75	65,00	65,00	66,25	2,17
49	66,25	68,75	65,00	66,67	1,91
56	68,75	68,75	66,25	67,92	1,44
60	68,75	68,75	66,25	67,92	1,44
63	67,50	63,75	63,75	65,00	2,17
66	67,50	66,25	63,75	65,83	1,91
73	67,50	65,00	63,75	65,42	1,91
79	67,50	65,00	66,25	66,25	1,25
85	68,75	65,00	66,25	66,67	1,91
90	67,50	65,00	66,25	66,25	1,25

Rep - repetition
P.D - Standard Deviation

	T4 (Flour	Rice)			
Days	Rep. 1	Rep. 2	Rep. 3	Average	D.P
2	93,75	93,75	93,75	93,75	0,00
4	93,75	92,50	93,75	93,33	0,72
7	93,75	93,75	93,75	93,75	0,00
9	93,75	92,50	93,75	93,33	0,72
11	88,75	88,75	88,75	88,75	0,00
14	87,50	87,50	87,50	87,50	0,00
15	85,00	87,50	87,50	86,67	1,44
17	87,50	87,50	87,50	87,50	0,00
21	86,25	87,50	87,50	87,08	0,72
23	87,50	87,50	87,50	87,50	0,00
28	87,50	87,50	87,50	87,50	0,00
30	87,50	87,50	87,50	87,50	0,00
31	87,50	87,50	87,50	87,50	0,00
32	87,50	87,50	87,50	87,50	0,00
36	87,50	87,50	87,50	87,50	0,00
42	87,50	87,50	87,50	87,50	0,00
49	87,50	87,50	87,50	87,50	0,00
56	87,50	87,50	87,50	87,50	0,00
60	87,50	87,50	87,50	87,50	0,00
63	87,50	87,50	88,75	87,92	0,72
66	87,50	87,50	87,50	87,50	0,00
73	87,50	87,50	87,50	87,50	0,00
79	88,75	88,75	87,50	88,33	0,72
85	87,50	87,50	87,50	87,50	0,00
90	87,50	87,50	87,50	87,50	0,00

Rep - repetition
P.D - Standard Deviation

APPENDIX H. Percentage of stabilized phase obtained in the spontaneous sedimentation tests of guava nectar with 0.05% xanthan gum + 0.05% guar gum, carried out over 90 days.

	T5 (Xanthan + Guar)				
Days	Rep. 1	Rep. 2	Rep. 3	Average	D.P
2	87,50	87,50	87,50	87,50	0,00
4	77,50	76,25	83,75	79,17	4,02
7	66,25	71,25	72,50	70,00	3,31
9	66,25	63,75	71,25	67,08	3,82
11	66,25	63,75	68,75	66,25	2,50
14	65,00	68,75	68,75	67,50	2,17
15	63,75	67,50	68,75	66,67	2,60
17	63,75	67,50	67,50	66,25	2,17
21	62,50	62,50	62,50	62,50	0,00
23	62,50	67,50	67,50	65,83	2,89
28	62,50	67,50	68,75	66,25	3,31
30	62,50	67,50	66,25	65,42	2,60
31	62,50	67,50	66,25	65,42	2,60
32	62,50	67,50	67,50	65,83	2,89
36	62,50	67,50	67,50	65,83	2,89
42	62,50	66,25	66,25	65,00	2,17
49	61,25	66,25	66,25	64,58	2,89
56	62,50	63,75	63,75	63,33	0,72
60	62,50	63,75	63,75	63,33	0,72
63	60,00	63,75	66,25	63,33	3,15
66	61,25	63,75	65,00	63,33	1,91
73	61,25	63,75	65,00	63,33	1,91
79	60,00	63,75	65,00	62,92	2,60
85	60,00	65,00	66,25	63,75	3,31
90	60,00	63,75	65,00	62,92	2,60

Rep - repetition
P.D - Standard Deviation

APPENDIX L. Percentage of stabilized phase obtained in the spontaneous sedimentation tests of guava nectar added with 1400 ppm of pectinase enzyme, carried out over 90 days.

Days	T6 (Xanthan + Rice Flour)				
	Rep. 1	Rep. 2	Rep. 3	Average	D.P
2	72,50	67,50	68,75	69,58	2,60
4	66,25	61,25	62,50	63,33	2,60
7	63,75	56,25	56,25	58,75	4,33
9	63,75	65,00	62,50	63,75	1,25
11	62,50	58,75	62,50	61,25	2,17
14	61,25	57,50	62,50	60,42	2,60
15	61,25	56,25	58,75	58,75	2,50
17	61,25	56,25	56,25	57,92	2,89
21	60,00	56,25	58,75	58,33	1,91
23	60,00	56,25	60,00	58,75	2,17
28	60,00	56,25	61,25	59,17	2,60
30	60,00	56,25	60,00	58,75	2,17
31	60,00	56,25	60,00	58,75	2,17
32	60,00	55,00	60,00	58,33	2,89
36	60,00	56,25	60,00	58,75	2,17
42	60,00	56,25	60,00	58,75	2,17
49	60,00	60,00	61,25	60,42	0,72
56	58,75	55,00	58,75	57,50	2,17
60	58,75	55,00	58,75	57,50	2,17
63	58,75	58,75	60,00	59,17	0,72
66	58,75	56,25	58,75	57,92	1,44
73	58,75	56,25	57,50	57,50	1,25
79	57,50	56,25	58,75	57,50	1,25
85	57,50	56,25	56,25	56,67	0,72
90	57,50	56,25	56,25	56,67	0,72

Rep - repetition
P.D - Standard Deviation

APPENDIX J. Percentage of stabilized phase obtained in the spontaneous sedimentation tests of guava nectar with 0.05% guar gum + 0.05% rice flour, carried out over 90 days.

T7 (Guar + Farinha de Arroz)

Days	Rep. 1	Rep. 2	Rep. 3	Average	D.P
2	88,75	87,50	87,50	87,92	0,72
4	85,00	86,25	87,50	86,25	1,25
7	77,50	78,75	86,25	80,83	4,73
9	77,50	81,25	83,75	80,83	3,15
11	81,25	81,25	85,00	82,50	2,17
14	80,00	80,00	85,00	81,67	2,89
15	77,50	78,75	83,75	80,00	3,31
17	81,25	82,50	81,25	81,67	0,72
21	81,25	82,50	76,25	80,00	3,31
23	82,50	85,00	81,25	82,92	1,91
28	82,50	87,50	87,50	85,83	2,89
30	81,25	85,00	85,00	83,75	2,17
31	81,25	85,00	85,00	83,75	2,17
32	82,50	85,00	85,00	84,17	1,44
36	80,00	82,50	81,25	81,25	1,25
42	78,75	81,25	81,25	80,42	1,44
49	76,25	77,50	80,00	77,92	1,91
56	85,00	87,50	81,25	84,58	3,15
60	85,00	87,50	81,25	84,58	3,15
63	78,75	87,50	81,25	82,50	4,51
66	76,25	87,50	78,75	80,83	5,91
73	81,25	87,50	81,25	83,33	3,61
79	78,75	87,50	81,25	82,50	4,51
85	81,25	87,50	81,25	83,33	3,61
90	81,25	87,50	81,25	83,33	3,61

Rep - repetition
P.D - Standard Deviation

APPENDIX L. Percentage of stabilized phase obtained in the spontaneous sedimentation tests of guava nectar added with 1400 ppm of pectinase enzyme, carried out over 90 days.

T8 (En zima)					
Days	Rep. 1	Rep. 2	Rep. 3	Average	D.P
2	100,00	100,00	100,00	100,00	0,00
4	100,00	100,00	100,00	100,00	0,00
7	100,00	100,00	100,00	100,00	0,00
9	100,00	100,00	100,00	100,00	0,00
11	100,00	100,00	100,00	100,00	0,00
14	100,00	100,00	100,00	100,00	0,00
15	100,00	100,00	100,00	100,00	0,00
17	95,00	100,00	100,00	98,33	2,89
21	93,75	100,00	100,00	97,92	3,61
23	92,50	100,00	100,00	97,50	4,33
28	93,75	100,00	100,00	97,92	3,61
30	91,25	100,00	100,00	97,08	5,05
31	91,25	100,00	100,00	97,08	5,05
32	91,25	100,00	100,00	97,08	5,05
36	91,25	100,00	100,00	97,08	5,05
42	91,25	100,00	100,00	97,08	5,05
49	90,00	93,75	93,75	92,50	2,17
56	92,50	92,50	92,50	92,50	0,00
60	92,50	92,50	92,50	92,50	0,00
63	92,50	92,50	92,50	92,50	0,00
66	91,25	91,25	91,25	91,25	0,00
73	91,25	91,25	90,00	90,83	0,72
79	92,50	88,75	87,50	89,58	2,60
85	90,00	87,50	87,50	88,33	1,44
90	90,00	87,50	87,50	88,33	1,44

Rep - repetition
P.D - Standard Deviation

APPENDIX M. Percentage of stabilized phase obtained in the spontaneous sedimentation tests of guava nectar added with 1400 ppm of pectinase enzyme + 0.1 % xanthan gum, carried out over 90 days.

	T9 (Enzyme + Xanthan)				
Days	Rep. 1	Rep. 2	Rep. 3	Average	D.P
2	100,00	100,00	100,00	100,00	0,00
4	100,00	100,00	100,00	100,00	0,00
7	87,50	93,75	87,50	89,58	3,61
9	87,50	100,00	100,00	95,83	7,22
11	80,00	81,25	81,25	80,83	0,72
14	81,25	81,25	81,25	81,25	0,00
15	82,50	81,25	78,75	80,83	1,91
17	81,25	81,25	78,75	80,42	1,44
21	76,25	81,25	73,75	77,08	3,82
23	76,25	81,25	71,25	76,25	5,00
28	76,25	81,25	68,75	75,42	6,29
30	78,75	75,00	73,75	75,83	2,60
31	78,75	75,00	73,75	75,83	2,60
32	78,75	78,75	75,00	77,50	2,17
36	76,25	76,25	72,50	75,00	2,17
42	78,75	77,50	73,75	76,67	2,60
49	80,00	76,25	75,00	77,08	2,60
56	80,00	81,25	75,00	78,75	3,31
60	80,00	81,25	75,00	78,75	3,31
63	87,50	81,25	75,00	81,25	6,25
66	81,25	80,00	75,00	78,75	3,31
73	85,00	77,50	72,50	78,33	6,29
79	81,25	76,25	77,50	78,33	2,60
85	82,50	77,50	81,25	80,42	2,60
90	82,50	77,50	81,25	80,42	2,60

Rep - repetition
P.D - Standard Deviation

APPENDIX N. Percentage of stabilized phase obtained in the spontaneous sedimentation tests of guava nectar added with 1400 ppm of pectinase enzyme + 0.1 % guar gum, carried out over 90 days.

	T10 (Enzyme + Guar)				
Days	Rep. 1	Rep. 2	Rep. 3	Average	D.P
2	100,00	100,00	100,00	100,00	0,00
4	100,00	100,00	100,00	100,00	0,00
7	100,00	100,00	100,00	100,00	0,00
9	100,00	100,00	100,00	100,00	0,00
11	100,00	100,00	100,00	100,00	0,00
14	100,00	100,00	100,00	100,00	0,00
15	100,00	100,00	100,00	100,00	0,00
17	100,00	100,00	100,00	100,00	0,00
21	100,00	100,00	100,00	100,00	0,00
23	100,00	100,00	100,00	100,00	0,00
28	100,00	100,00	100,00	100,00	0,00
30	100,00	100,00	100,00	100,00	0,00
31	100,00	100,00	100,00	100,00	0,00
32	100,00	100,00	100,00	100,00	0,00
36	100,00	100,00	100,00	100,00	0,00
42	100,00	100,00	100,00	100,00	0,00
49	100,00	100,00	100,00	100,00	0,00
56	100,00	100,00	100,00	100,00	0,00
60	100,00	100,00	100,00	100,00	0,00
63	100,00	100,00	100,00	100,00	0,00
66	93,75	100,00	100,00	97,92	3,61
73	93,75	100,00	100,00	97,92	3,61
79	93,75	93,75	100,00	95,83	3,61
85	91,25	93,75	100,00	95,00	4,51
90	91,25	81,25	100,00	90,83	9,38

Rep - repetition
P.D - Standerd Deviation

APPENDIX O. Percentage of stabilized phase obtained in the spontaneous sedimentation tests of guava nectar with 1400 ppm pectinase enzyme + 0.1% rice flour, carried out over 90 days.

Days	T11 (Enzyme + Rice Flour)			Average	D.P
	Rep. 1	Rep. 2	Rep. 3		
2	100,00	100,00	100,00	100,00	0,00
4	100,00	100,00	100,00	100,00	0,00
7	91,25	93,75	93,75	92,92	1,44
9	87,50	90,00	90,00	89,17	1,44
11	81,25	85,00	87,50	84,58	3,15
14	81,25	76,25	83,75	80,42	3,82
15	80,00	76,25	85,00	80,42	4,39
17	80,00	76,25	81,25	79,17	2,60
21	76,25	75,00	76,25	75,83	0,72
23	80,00	77,50	81,25	79,58	1,91
28	77,50	75,00	80,00	77,50	2,50
30	77,50	75,00	77,50	76,67	1,44
31	77,50	75,00	77,50	76,67	1,44
32	77,50	76,25	77,50	77,08	0,72
36	76,25	75,00	76,25	75,83	0,72
42	76,25	75,00	75,00	75,42	0,72
49	76,25	75,00	75,00	75,42	0,72
56	76,25	75,00	76,25	75,83	0,72
60	76,25	75,00	75,00	75,42	0,72
63	76,25	75,00	73,75	75,00	1,25
66	76,25	75,00	73,75	75,00	1,25
73	76,25	75,00	75,00	75,42	0,72
79	76,25	75,00	75,00	75,42	0,72
85	76,25	75,00	73,75	75,00	1,25
90	76,25	75,00	73,75	75,00	1,25

Rep - repetition
P.D - Standard Deviation

APPENDIX P. Percentage of stabilized phase after 90 days of storage

Stabilized Phase (%)		
Days T1 T2 T3 T4 T5 T6 T7 T8	T9	T10T11
90 74 b 64 c 66 c 87 a 63 c 57 d 83 a 88 a	80 ab	91 a 75 b
- Averages followed by equal letters do not differ by Tukey's test (p< 0.05).	.	
- The averages shown correspond to the average of the three readings at 90 days of storage T1-control, T2-xanthan 0.1%, T3-guar 0.1%, T4-rice flour 0.1%, T5-xanthan 0.05% + guar		
0.05%, T6-xanthan 0.05% + rice flour 0.05%, T7-guar 0.05% +	flour	rice 0.05%, T8-
enzyme 1400 ppm, T9-enzyme 1400 ppm + xanthan 0.1%, T10-enzyme 1400 ppm + rice flour 0.1%.	1400 ppm +	guar 0.1%, T11-

yes

I want morebooks!

Buy your books fast and straightforward online - at one of world's fastest growing online book stores! Environmentally sound due to Print-on-Demand technologies.

Buy your books online at
www.morebooks.shop

Kaufen Sie Ihre Bücher schnell und unkompliziert online – auf einer der am schnellsten wachsenden Buchhandelsplattformen weltweit! Dank Print-On-Demand umwelt- und ressourcenschonend produziert.

Bücher schneller online kaufen
www.morebooks.shop

Printed by Books on Demand GmbH, Norderstedt / Germany